Lecture Notes
in Physics

Edited by H. Araki, Kyoto, J. Ehlers, München, K. Hepp, Zürich
R. Kippenhahn, München, H. A. Weidenmüller, Heidelberg
and J. Zittartz, Köln
Managing Editor: W. Beiglböck

258

Wm. G. Hoover

Molecular Dynamics

Springer-Verlag

Berlin Heidelberg New York London Paris Tokyo

7332/011

Author

William Graham Hoover
University of California, College of Engineering
Department of Applied Science Davis-Livermore
and Physics Department, Lawrence Livermore National Laboratory
Hertz Hall, P. O. Box 808, L-794, Livermore, CA 94550, USA

ISBN 3-540-16789-7 Springer-Verlag Berlin Heidelberg New York
ISBN 0-387-16789-7 Springer-Verlag New York Berlin Heidelberg

Printing and binding: Beltz Offsetdruck, Hemsbach/Bergstr.
2153/3140-543210

Preface:

The University of Vienna was founded in 1365 and now has about 48 000 students. Here Boltzmann began and ended his scientific career, as mathematician, physicist, and natural philosopher. His main research interest was understanding the microscopic basis of macroscopic thermodynamic and hydrodynamic phenomena. Boltzmann was specially interested in irreversibility—in understanding how the formally reversible equations of mechanics can give rise to the blatant irreversibility summarized by the Navier-Stokes equations and the second law of thermodynamics.

Boltzmann's heritage is nurtured in Vienna. Research into the instabilities exhibited by chaotic dynamical systems and the correlations between molecules, as revealed by scattering and described by statistical dynamical equations, is still being actively pursued here. It is particularly appropriate to describe recent developments in numerical kinetic theory, "molecular dynamics", at Boltzmann's own University. Complementary developments in computer technology and in the structure of nonequilibrium dynamics itself have helped to bring Boltzmann's goal of understanding irreversibility closer to realization during the past decade. It seems likely that these technical advances in nonequilibrium molecular dynamics will be useful in suggesting new methods for treating and understanding quantum mechanical systems. Atomistic simulation is developing rapidly on many fronts, as evidenced by the many international schools of physics and workshops devoted to this topic. The Paris workshops organized by Carl Moser at CECAM (Centre Européen de Calcul Atomique et Moléculaire) have been particularly useful and stimulating.

These notes cover a series of lectures delivered to graduate students and faculty at the University of Vienna during the summer semester in 1985. Those attending had rather varied backgrounds, but were well-grounded in thermodynamics and equilibrium statistical mechanics. The lectures summarized the current state of what is now a world-wide activity, molecular dynamics simulations. This summary includes a variety of sample problems taken from the literature. Because I was invited to lecture on my research interests, the approach followed emphasizes physical and intuitive arguments at the expense of mathematical ones. For me, the path of least resistance was to illustrate these arguments with sample problems from my own work. For this same reason simple systems, with short-ranged forces and only a few degrees of freedom, predominate among the examples chosen. These notes can serve to introduce the subject of molecular dynamics at either the senior-year undergraduate or the graduate level as well as to stimulate new developments along the rapidly moving research frontier. I have made no effort to attain comprehensive coverage. But the reader will have no difficulty in finding more examples, by the hundreds, in the rapidly expanding research literature on molecular dynamics.

Dr. Karl Kratky and Professor Dr. Peter Weinzierl were both helpful in arranging for this visit and in providing gemütlich living and working conditions. Drs. Harald Posch and Franz Vesely kindly offered comments on the oral version of these notes. Dr. William T. Ashurst provided useful criticism of the typed draft. Mike Pound and Belen Flores assisted with the intricacies of computer-assisted type setting, using Donald Knuth's T_EX system. I thank them, as well as the Universities of Vienna and California for support during this period. I prepared the final manuscript version under the auspices of the United States Department of Energy, at the Lawrence Livermore National Laboratory in California through Contract W-7405-Eng. I thank Professor Fred Wooten, Chair of the Department of Applied Science at the University of California at Davis/Livermore, and Dr. Lewis Glenn, Leader of the Theoretical and Applied Mechanics Group at the University of California's Lawrence Livermore National Laboratory, for encouraging and supporting the publication of these notes. Dr. Giulia De Lorenzi was particularly helpful in criticizing obscurities in the notes and assisting with some of the Figures. These notes are dedicated to her.

Vienna, March-June 1985

Livermore, May-June 1986

TABLE OF CONTENTS

I. Historical Development and Scope

    A. Newton's Mechanics 1

    B. Lagrange's Mechanics and Hamilton's Least Action Principle 13

    C. Hamilton's Mechanics - Introduction to Nosé's Mechanics 16

    D. Gauss' Mechanics and the Principle of Least Constraint 22

    E. Nosé's Mechanics - Temperature and Pressure Constraints 27

    F. Numerical Mechanics - Fermi, Alder, Vineyard, and Rahman 35

Bibliography for Chapter I. 41

II. Connecting Molecular Dynamics to Thermodynamics

    A. Instantaneous Mechanical Variables 42

    B. Macroscopic Dynamics 47

    C. Virial Theorem and Heat Theorem 54

    D. Elastic Constants 59

    E. Number-Dependence 62

    F. Results 67

Bibliography for Chapter II. 70

III. Newtonian Molecular Dynamics for Nonequilibrium Systems

    A. Limitations of the Newtonian Approach 72

    B. Gases: Boltzmann's Equation 72

    C. Liquids: Shockwave Simulation and Fragmentation 77

    D. Solids: Breakdown of Continuum Mechanics 84

Bibliography for Chapter III. 91

IV. Nonequilibrium Molecular Dynamics

    A. Motivation for Generalizing Newton's Equations of Motion 92

    B. Control Theory and Feedback 94

    C. Examples of Control Theory: "Isothermal" Molecular Dynamics 94

    D. Heat Conducting Chain - 9 Examples 97

    E. Linear and Nonlinear Response Theory 101

    F. Diffusion and Viscosity for Two Hard Disks 102

    G. Simulation of Diffusive Flows 106

    H. Simulation of Viscous and Plastic Flows 114

    I. Simulation of Heat Flows 124

Bibliography for Chapter IV. 131

V. Future Work

....132

Bibliography for Chapter V. 133

Index 134

I. HISTORICAL DEVELOPMENT AND SCOPE

I.A Newton's Mechanics

Sir Isaac Newton (1642-1727) developed mechanics in order to understand the motion of the planets and their moons. He recognized that the same mechanical laws and the same "universal" gravitational attraction apply to all bodies, large and small. This correlation, spanning length and time scales from astronomical to atomic, set the stage for the study of smaller-scale terrestrial problems in fluid and solid mechanics, ranging down to those of special interest to Boltzmann and to us, the "molecular dynamics" of atomistic particles. That these problems are many, important, and varied can be verified by skimming any of the dozens of research journals devoted to their solutions.

There are still interesting conceptual problems associated with applying Newton's mechanics in practice: how can the many degrees of freedom in a macroscopic system be described most simply? That is, how does the few-parameter description of thermodynamics and hydrodynamics come from the many-parameter microscopic description? How does the irreversible character of the macroscopic linear diffusion equation arise? How can we understand nonlinear phenomena which lie far from equilibrium without unnecessary complexity? Atomistic molecular dynamics is an extremely useful tool in addressing all of these questions. It can reveal the hidden mechanisms and correlations which underly macroscopic behavior and it can contribute to the testing and improvement of theoretical descriptions.

All of the problems mechanics treats are idealized. They are not exact copies of nature, but are rather intended to illustrate the interesting features of nature in a relatively clear and simple way. Unimportant features are consciously omitted. Consider gravity. In a mole of material the gravitational accelerations affect the trajectories only beyond the tenth decimal place. Likewise, faraway stars and minor mountain ranges have a negligible influence on the evolution of the solar system. In atomistic mechanics our model of the atoms themselves is imperfect, not so much because quantum mechanics is ignored, but because our knowledge of constitutive behavior is still inadequate for the reliable construction of interatomic forces that can reproduce the experimental data.

Because the best available trajectory data were those establishing Kepler's laws of planetary motion, the proving ground for Newton's ideas was astronomical. Kepler found that [1] the planets travel in elliptical orbits with the sun at foci of these ellipses, that [2] the vector joining the moving planet to the sun sweeps out area at a constant rate, and that [3] the period during which each ellipse is traced out varies as the three-halves power of the orbit's size. In **Figure 1** we include finite-difference solutions of Newton's equations of motion for two different orbits. The orbits were generated using the "Verlet algorithm" described in more detail below. The orbits are shown as series of dots to emphasize their finite-difference source.

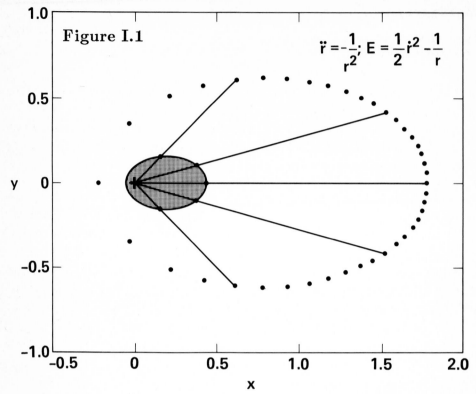

Figure I.1

$$\ddot{r} = -\frac{1}{r^2}; \; E = \frac{1}{2}\dot{r}^2 - \frac{1}{r}$$

The orbits serve to illustrate all three of Kepler's laws. Both orbits are ellipses, composed of pie-slice pieces of equal areas. The relative size of the orbits has been chosen so that the period of the larger one is exactly eight times that of the smaller one. In accordance with Kepler's third law the ratio of the larger orbit's major axis size $2R$ to that of the smaller orbit $2r$ is $8^{2/3} = 4$.

The last two of Kepler's three laws are easily seen as consequences of the conservation of angular momentum and the virial theorem, respectively. As an introduction to mechanics, helping to establish the notation we use for more complicated many-body systems, let's consider these laws in more detail. For convenience, we choose a plane polar-coordinate system with an infinitely-massive point, the sun, at the origin.

The angular momentum of a particle, or planet, of mass m travelling around the sun, is given by the cross-product of the vectors mr and v, where the velocity v is not necessarily normal to the direction of r. The magnitude of the cross-product $mr \times v$ is also $2m$ times the rate at which the radius vector sweeps out area in the plane of the orbit. This equivalence of angular momentum with the rate of area generation establishes Kepler's second law.

Kepler's third law is simply related to the "virial theorem", a theorem to which we will return in Section C of Chapter II, in greater generality. To derive the simplest, scalar, version of this theorem we can take the dot product of the radius vector r and Newton's equation of motion for the particle of mass m located at r:

$$m\ddot{r} = F.$$

The righthand side of the dot product, $r \cdot F$, is equal to the gravitational potential. A substitution,

$$(mr\ddot{r}) = md(r\dot{r})/dt - m\dot{r}^2,$$

followed by a time-average over one full orbit, shows that the gravitational potential energy is also equal to minus twice the kinetic energy. This equivalence, between the potential, which depends on the orbital radius as $[1/r]$, and minus twice the kinetic energy, which depends on the radius r and the period τ as $[r^2/\tau^2]$, establishes Kepler's third law. Alternatively, this law can be used to infer the inverse-square form of the acceleration due to gravity.

In astronomical problems the accelerations are gravitational. In the atomistic problems of molecular dynamics the accelerations are much shorter ranged, acting over distances of the order of nanometers. In truth, atomistic systems are governed by quantum mechanics, but in these notes we mainly ignore quantum effects, concentrating almost completely on classical mechanics and its recent modifications. The rationale for this choice lies in the fact that the qualitative features of thermodynamic and hydrodynamic behavior are scarcely affected by quantum mechanics, while the difficulties involved in implementing quantum simulations, especially away from equilibrium, are formidable. Thus we will generate atomistic trajectories using techniques borrowed from macroscopic Newtonian mechanics. Newtonian mechanics is concerned with using the "accelerations" which result when "forces" act on "masses". The goal of Newtonian mechanics is the calculation of the velocities and trajectories of these masses.

In the absence of "accelerations" inertial-frame particles move along straight-line trajectories. Changes in this straight-line behavior are due to "forces". Particles respond to the forces imposed upon them by undergoing accelerations inversely proportional to their "masses". If these accelerations are given functions of the coordinates, such as the gravitational accelerations acting among the planets and the stars or the idealized hard-sphere repulsions acting between billiard-ball caricatures of atoms, the corresponding velocity changes can be calculated. Newton's second-order equations of motion cannot be solved without first specifying appropriate boundary conditions. In the usual case these would be the initial values of the coordinates and velocities. Once these initial values are given the system of equations $m\ddot{r} = F(r)$ can be solved by straightforward numerical methods.

In his CalTech lectures, Feynman provides two nice illustrations to introduce numerical simulation techniques. He works out two elliptical trajectories of the type shown in **Figure 1** illustrating Kepler's laws, the phase-space trajectory of a harmonic oscillator and the coordinate-space trajectory of a planet about the sun. In both calculations he uses a centered second-difference Störmer algorithm which has come to be known as the "Verlet algorithm":

$$m\big[r(t + dt) - 2r(t) + r(t - dt)\big] = F(t)\,dt^2.$$

In order to start the algorithm it is necessary to *approximate* the initial velocity by either $[r(0) - r(-dt)]/dt$ or $[r(dt) - r(-dt)]/(2dt)$. Given the coordinates at times 0 and $-dt$ the Störmer-Verlet finite-difference equation can be solved for new values at times $+dt, +2dt, \ldots$. It is easy to see (by expanding the left hand side in powers of dt) that the difference equation becomes exact through terms of order dt^3 for small dt. Because the "local" single-step error in $r(t + dt)$ is of order dt^4 it might appear that the "global" long-time error at time τ, after $[\tau/dt]$ steps, would vanish as dt^3. But the error is actually larger, of order dt^2, because the equation of motion is *second order*.

The *reversibility* of Newtonian mechanics is retained by Verlet's approximate "finite-difference" representation. "Reversibility" means that a movie of the motion, run backward, would satisfy exactly the same equations of motion. There is nothing in the mathematically symmetric structure of the equations to suggest that they really do produce physically irreversible behavior.

On the other hand, thermodynamics is certainly an accurate description of material behavior. And thermodynamics is not reversible. In isolated systems the *energy* is fixed (first law of thermodynamics) but the *entropy* behaves irreversibly (second law of thermodynamics), always increasing with time in isolated systems not in complete thermal, mechanical, and chemical equilibrium. For more than 100 years there has been a continuous discussion of how reversible equations can produce, or be used to explain, irreversible behavior.

The apparent contradiction between the microscopic and macroscopic viewpoints can be concisely expressed in terms of the Zermelo-Poincaré and Loschmidt objections to the explanation of irreversibility using reversible mechanics:

(ZP) for an isoenergetic Hamiltonian system with a bounded "phase space" the long-time solution of Newton's equations must eventually approach the initial conditions arbitrarily closely (Zermelo-Poincaré recurrence). [Phase space is the $2f$-dimensional space whose orthogonal axes correspond to the f coordinates and f momenta required to describe the state of a system with f degrees of freedom.]

(L) the reversibility of Newton's equations means that any entropy-producing trajectory could be reversed in time to make an equally valid trajectory along which the entropy would decrease (Loschmidt objection).

Both objections can be illustrated by the harmonic oscillator trajectories shown in **Figure 2**. Choosing an oscillator with unit mass and force constant gives an oscillator period of 2π. The recurrence time, for *any* initial state, is 2π. **Figure 2** shows also the reversibility of oscillator trajectories. The q and $\dot{q}$ histories shown cover $7/12$ of an oscillator period in the forward direction, according to the equations $q = \sin(t)$ and $\dot{q} = \cos(t)$. In the reversed direction $\dot{q}$ changes sign, and, after $7/12$ of an oscillator period, the oscillator returns to its initial state. Thus an

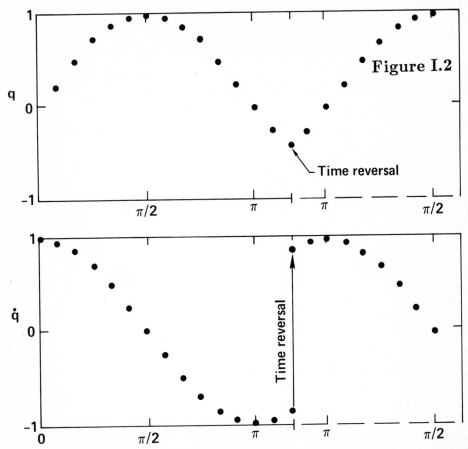

Figure I.2

oscillator illustrates both the Zermelo-Poincaré recurrence and the Loschmidt reversibility alleged to contradict the second law of thermodynamics. But the two objections to the use of reversible mechanics as the basis for understanding irreversible phenomena can be countered in several ways:

(i) the time required for Poincaré recurrence is outrageously large, reaching the age of the universe for systems with only a few dozen degrees of freedom.

(ii) the information required to produce a reversed set of trajectories is not available experimentally.

(iii) the macroscopic equations refer only to "average" behavior. [Averages are constructed by including many similar systems which begin with different initial conditions.] The fluctuations seen in small systems can violate the second law of thermodynamics.

(iv) the influence of boundaries, faraway stars, Coriolis accelerations, in fact, *any* time-varying influence from outside the system, effectively destroys reversibility.

(v) the Lyapunov instability of the equations (discussed below) makes it impossible in principle to solve them for long times.

(vi) only very simple systems such as the harmonic oscillator whose coordinate and velocity trajectories are shown in **Figure 2** are reversible. Systems only a little more complicated than the oscillator typically exhibit the chaotic irreversible behavior characteristic of real materials.

The discussion of reversibility may well continue for another 100 years, but it appears likely that it is primarily a mathematical, as opposed to physical, problem. The intuitive ideas of Boltzmann, which link reversible motion equations to irreversibility, have been buttressed by the computer calculations carried out since the second World War. There is now no reasonable doubt that the solutions of the reversible equations exhibit the *same* irreversible behavior that thermodynamics and fluid dynamics describe for macroscopic systems. Furthermore, the study of the solutions of Newton's reversible equations, using molecular dynamics, reveals the mechanism for that irreversible behavior, the Lyapunov instability of the underlying equations.

Molecular Dynamics calculations are not unduly difficult. Feynman obtains three-digit accuracy in both of his elliptic-orbit back-of-the-envelope calculations. If such a calculation involves more than a few time steps or more than a few degrees of freedom it is worthwhile to use a fast computer to do the work. Typical molecular dynamics simulations involve 32, 108, 256, ... particles (numbers chosen to be consistent with periodic face-centered cubic packing) and would be far from fast if pursued with pencil and paper. "Fast" is just fast relative to human speed, and includes the local departmental VAX computer as well as the dozen or so "multimegaflop" ("flops" is an acronym for *floating-point operations per second*) computers a hundred times faster that are used at the Lawrence Livermore Laboratory.

There is no reasonable doubt that very high accuracy is unnecessary in solving Newton's equations of motion unless one is specially interested in *rigor mortis* properties related to mathematical reversibility. High accuracy might prove necessary in classical periodic orbit calculations to be used as bases for estimating semi-classical quantum properties. But in a typical study intended to measure the pressure of a many-body system within one percent, one would expect to need on the order of only six-digit accuracy in the fundamental description of the dynamics. The dynamics must be known with a slightly greater accuracy than the desired thermodynamic and hydrodynamic properties because the contributing variables are generated by a differential equation. Thus the velocities and the accelerations (first and second time derivatives of the particle trajectories) contribute to the energy, pressure, and other macroscopic properties of interest. Not just the trajectory, but also its first and second time derivatives need to be given with an accuracy comparable to that of the desired macroscopic averages.

The trajectory equations for most nonlinear problems exhibit what is called "Lyapunov instability". This means that the separation in the phase space between two neighboring phase-space trajectories increases, exponentially fast, on the average, with time. The separation continues to increase until it approaches a value imposed by the geometric and energetic constraints on the system.

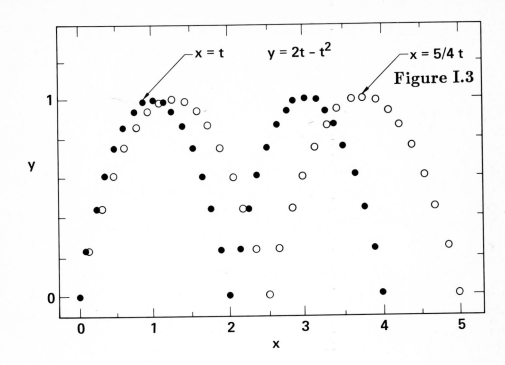

Figure I.3

We illustrate the non-Lyapunov *stable* case in **Figure 3** by showing the motion of a particle of unit mass in a gravitational field. The filled circles correspond to the trajectory of such a particle moving to the right, at unit x velocity, while bouncing vertically on the $y = 0$ plane in a gravitational field with $g = 2$. A similar particle, with x velocity component 1.25 rather than 1.00, traces out the trajectory indicated by open circles. The two trajectories depart from each other *linearly*, not exponentially, with time. The motion is therefore "stable".

We illustrate the Lyapunov *unstable* case by showing two representations (linear in **Figure 4** and semilogarithmic in **Figure 5**) of a ball bouncing on a unit sphere, this time with a gravitational field of unity. The motion of this system is identical to that in which a ball with unit diameter bounces on a fixed sphere of the same size. Initially the bouncing mass point is offset by 0.00001 diameters relative to the fixed lower ball. After nine bounces this offset has increased to exceed the radius of the lower ball. The exponential growth of the offset is emphasized in **Figure 5** by plotting the logarithm of the offset, which grows very nearly linearly with the number of bounces. Each bounce increases the offset by about a factor of 3.7.

Boltzmann understood this dynamical instability, but widespread recognition that it is inherent in *most* systems of nonlinear differential equations was long in coming. The instability is easily illustrated by calculating the maximum number of bounces made by an idealized, perfectly elastic and perfectly round, billiard ball, dropped from above on to a similar ball.

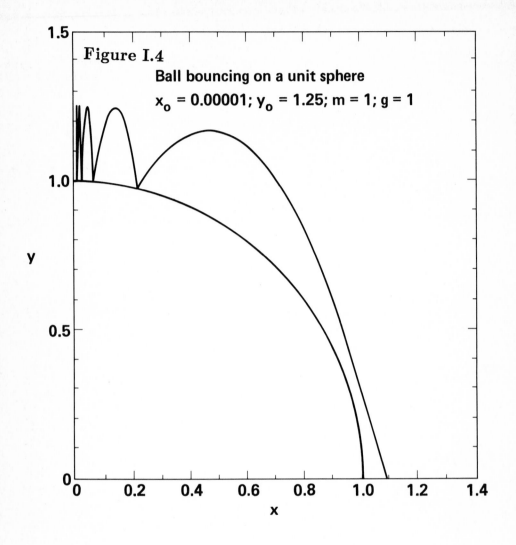

Figure I.4
Ball bouncing on a unit sphere
$x_0 = 0.00001$; $y_0 = 1.25$; $m = 1$; $g = 1$

It is 9 in the case shown in **Figure 4** above and in **Figure 5** with an offset of 0.00001. In principle, if the dropping ball were exactly centered over the lower ball (and perfectly elastic) the motion would continue, periodically, forever. On the other hand, a computer simulation of the experiment would be likely to predict different results. The finite precision of the computer calculation (typically between 8 and 14 decimal digits) would result in the upper ball's landing some distance away from the top of the lower ball, inducing a horizontal acceleration and velocity. Both the acceleration and the velocity would then act to increase the offset on the second bounce, roughly by a factor of ten, and then the offset would continue to increase exponentially until the upper ball had missed the lower one completely. There is no doubt that a careful attempt to carry out such an experiment in the laboratory would lead to the same kind of unstable behavior

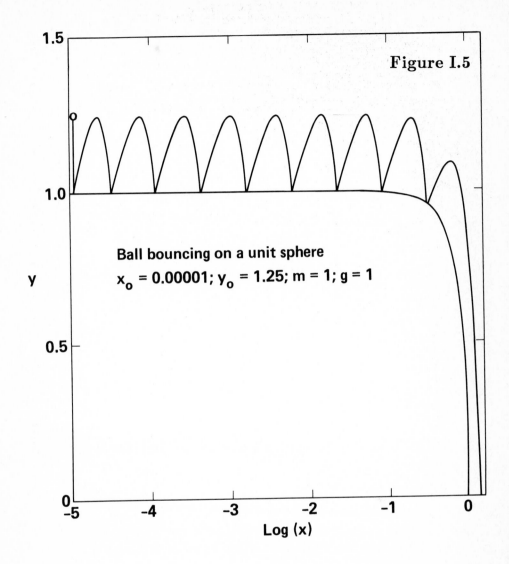

Figure I.5

Ball bouncing on a unit sphere
$x_o = 0.00001$; $y_o = 1.25$; $m = 1$; $g = 1$

y

Log (x)

seen in the computer experiment, but probably with many fewer bounces due to the relatively larger experimental asymmetries from nonuniformity and nonelasticity in real balls.

If we were to use quantum mechanics in setting up the initial conditions for the experiment much the same result would be obtained, even for idealized perfect spherical elastic balls. Consider Heisenberg's uncertainty principle. This principle places theoretical limits on the accuracy with which coordinates and momenta can simultaneously be known. The product of the two "uncertainties" is at least of order Planck's constant h. Thus the limited accuracy with which one ball can be centered over another (with the product $dp \times dq$ of order h) allows only (about) 17 bounces. Either real irregularities or the real space-momentum correlations described by quantum mechanics would lead to the unstable behavior seen in the idealized dropping-ball experiment.

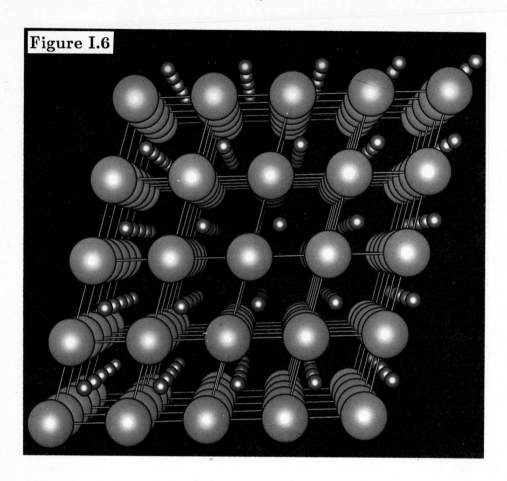

Figure I.6

Solutions of differential equations for dynamical systems require appropriate initial and boundary conditions. The problems with which Newton began had the simplest possible boundary, a perfect vacuum. In the strongly nonequilibrium problems to which molecular dynamics is now being applied such a simple vacuum boundary is not usually appropriate. Instead, the container must be taken into account, or explicitly avoided through the use of periodic boundary conditions. Periodic boundaries are illustrated in **Figure 6** for a three-dimensional two-body system by showing $125 = 5^3$ separate images of both particles. Periodic boundaries have the unusual property that they are inconsistent with the conservation of angular momentum. For example, **Figure 7** shows a particle passing out of the "top" of a periodic two-dimensional $L \times L$ system, with $y = +L/2$. The *sign* of y, and hence of $my\dot{x}$ changes discontinuously when the particle reenters at the bottom of the system, with $y = -L/2$. Thus one of the two contributions to the angular momentum, $p_\theta = m(x\dot{y} - y\dot{x})$ changes sign. The lack of conservation of angular momentum is not usually a problem, but this example points out the need for considering the effect of boundary conditions on observed properties.

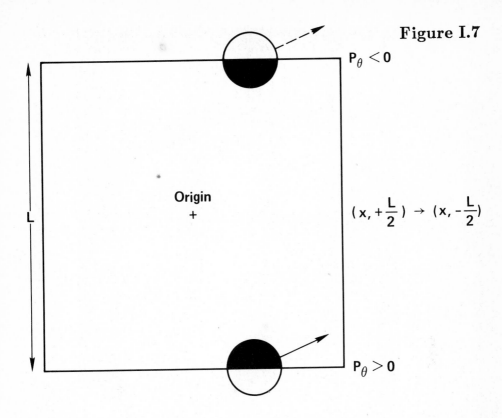

Figure I.7

$P_\theta < 0$

Origin
+

$\left(x, +\dfrac{L}{2} \right) \rightarrow \left(x, -\dfrac{L}{2} \right)$

$P_\theta > 0$

To illustrate simultaneously the ideas of Lyapunov instability and periodic boundaries, we show in the **Figure 8** three views of 125 superposed periodic 2-body systems undergoing shear deformation. The larger of the two particles forms a lattice undergoing steady shear. 125 independent images of the smaller particle are shown in each of the three snapshots. Initially the space coordinates of the small particles differed in the fourth digit only. These small differences led, after just a few collisions of the small particle with the large (but periodic) scatterer particle, to the nearly random small-particle distribution seen at the lower left of **Figure 8**.

Periodic boundaries have confused many students. Newton's small-particle acceleration equations, $m\ddot{r} = F$, can contain either an explicit vector sum of forces from *all* periodic images of the large particle or, if the coordinates of the smaller particle are replaced within the central box whenever a periodic boundary is crossed, and if the forces are short-ranged, only a few terms. The simplest procedure, exact for forces which vanish beyond half the periodic box width, is to select only the *nearest* particle for calculating the force. This is the "nearest-image" convention. For the short-ranged forces characteristic of neutral matter it is usual to include all particles lying within about three particle diameters in the force sum.

If gravitational forces were to be included, an attempt to work out the energy of an infinite array of identical unit cells would yield not only a divergent sum, but also a divergent sum for the interaction of each cell with the rest. So we have another good reason for ignoring gravity.

Figure I.8

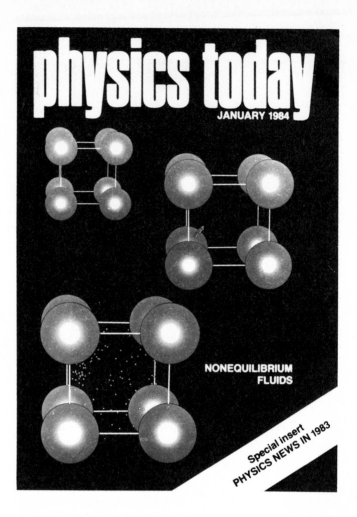

Gravity is inconsistent with the so-called "thermodynamic limit", in which the intensive properties converge to values given by the thermodynamic "infinite-system" equation of state.

To summarize, Newtonian mechanics consists of three elements:

(i) a recipe (forces F and masses m) giving the accelerations in terms of the coordinates r: $\ddot{r}(r) = F(r)/m$. Newton's recipe is a set of *second*-order ordinary differential equations.

(ii) boundary conditions for the differential equations, including both initial conditions and time-dependent boundary accelerations exerted on the system by the outside world.

(iii) an algorithm for solving the differential equations. Adams-Moulton, Runge-Kutta, or Gear methods can solve first-order equations such as Hamilton's, or the nonequilibrium equations of motion described in Chapter IV. Runge-Kutta algorithms are easily programmed and suitable for use on small personal computers.

I.B Lagrange's Mechanics and Hamilton's Least Action Principle

Joseph Louis Lagrange(1736-1813) studied in his native Italy and spent his research career in Berlin and Paris. He contributed to the description of physical systems by developing the theory of differential equations.

For conservative (constant-energy) systems, Lagrange's formulation of mechanics is specially useful in dealing with systems incorporating "holonomic" constraints. Taken literally, *holonomic* means "whole laws". In mechanics holonomic constraints involve geometric, as opposed to kinetic, constraints. Lagrange's equations of motion differ from Newton's in making the motion of constrained systems easier to treat. This is done by formulating equations of motion in terms of "generalized" coordinates $q(r, \dot{r})$, which can be complicated functions of the space coordinates r. In either the Newtonian case or the Lagrangian case, the coordinates and their time derivatives appear in equations for the accelerations.

How does Lagrangian mechanics work? The fundamental equation, the equation of motion, can be obtained from Hamilton's Principle, often, but not always, also called the *Principle of Least Action*. To start out, first write down the "Lagrangian", $L(m, q, \dot{q})$. The Lagrangian is the difference between the Kinetic Energy of the system, $K(m, q, \dot{q})$, written in terms of the masses m, generalized coordinates q, and velocities $\dot{q}$, and the Potential Energy, $\Phi : L = K - \Phi$. The kinetic energy is $mv^2/2$ summed over all mass elements. Hamilton's Least Action Principle then states that the motion $q(t)$ between any two fixed space-time points $q_0(t_0)$ and $q_1(t_1)$ is that which minimizes the integral of L relative to slightly different, varied motions:

$$\int L\,dt \quad \text{minimum,} \quad \text{or} \quad \int \delta L\,dt = 0.$$

Figure 9 illustrates this principle for a unit mass travelling from $x = 0$ to $x = 2$ in a gravitational field, with field energy $\phi = 2y$. We work out the action integral for three different parabolic trajectories, with the parameter α equal to 1, 2, and 3, as shown in **Figure 9**:

$$x = t; \quad y = \alpha t\,(1 - 0.5\,t).$$

For this set of trajectories the kinetic energy integrated over time is $1 + (1/3)\alpha^2$ and the potential energy integrated over time is $(4/3)\alpha$. The difference between these integrals is the *action integral*, $1 - (4/3)\alpha + (1/3)\alpha^2$, which has a minimum value, $-1/3$, for $\alpha = 2$. This path, which corresponds to the parabola satisfying the classical equations of motion, has a smaller action integral than do any neighboring paths. For instance, $\alpha = 1$ and $\alpha = 3$ correspond to greater, and therefore unacceptable, action integrals of 0. But there is a fly in the ointment. A path which passes *through* the barrier shown in **Figure 9** can have an arbitrarily large negative action (by making the potential barrier, shown as a rectangular box, arbitrarily large and positive). This example emphasizes the local nature of the variation in Hamilton's principle. The principle must be applied locally to avoid drawing qualitatively incorrect conclusions.

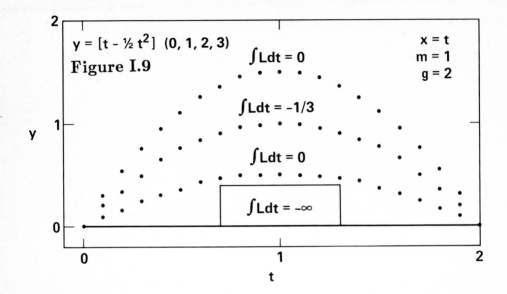

The figure contains: $y = [t - \tfrac{1}{2} t^2]$ (0, 1, 2, 3), Figure I.9, $\int L\,dt = 0$, $\int L\,dt = -1/3$, $\int L\,dt = 0$, $\int L\,dt = -\infty$, $x = t$, $m = 1$, $g = 2$

Because, in the general case, L depends explicitly upon both q and its time derivative $\dot{q}$, the variation has the form

$$\int \delta L\,dt = 0 = \int \big[(\partial L/\partial q)\,\delta q + (\partial L/\partial \dot{q})\,(d\,\delta q/dt)\big]\,dt.$$

The derivative $(d\,\delta q/dt)$ can be integrated (with respect to time) by parts, because the integral, δq, is known to vanish at the two endpoints. Then, because the form of the variation that results, $\int\big[(\partial L/\partial q) - (d/dt)(\partial L/\partial \dot{q})\big]\,\delta q\,dt$ contains the *arbitrary* variation δq, the coefficient of δq must vanish at all times in the interval. This requirement establishes Lagrange's equations of motion,

$$(d/dt)\,(\partial L/\partial \dot{q}) = \partial L/\partial q.$$

Because the Lagrangian equations of motion still don't convey much information, and are identical to Newton's equation $m\ddot{r} = F$ provided that $\partial L/\partial \dot{q}$ corresponds to $m\dot{q}$ and $\partial L/\partial q$ corresponds to F, we consider an example with constraints. Shown in **Figure 10** is a two-dimensional system involving two holonomic constraints, a triatomic molecule with Particles labelled 0, 1, and 2, with the two distances r_{10} and r_{20} fixed at 1. We choose masses for the particles which make the problem as simple as possible: the central particle, Particle 0, has no mass; Particles 1 and 2 have equal masses, m. Because the center-of-mass motion of $(r_1 + r_2)/2$ has no bearing on the motion relative to the center of mass, we consider the case in which the center of mass is fixed at the origin.

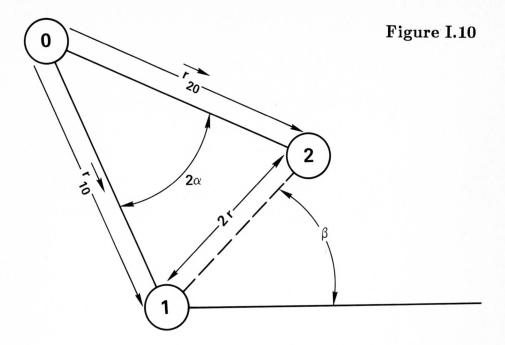

Figure I.10

In the Newtonian mechanics of Section A of Chapter I or the Gaussian mechanics to be described in Section D of Chapter I we would need to introduce "constraint" forces to keep these two distances fixed. In Lagrangian mechanics we can avoid defining the constraint forces by incorporating the constraints in our description of the problem, using as "generalized coordinates" α, half the angle between the 01 and 02 directions, and the angle β between r_{12} and some fixed axis in space, chosen horizontal in **Figure 10**. The variable r is equal to $\sin\alpha$.

Because there is no potential energy, the Lagrangian and the kinetic energy K are identical for this problem:

$$L = K - \Phi = K = m\left[(\dot{\beta}\sin\alpha)^2 + (\dot{\alpha}\cos\alpha)^2\right].$$

From this Lagrangian, to which we return in the following section on Hamiltonian mechanics, $\alpha(t)$ and $\beta(t)$ can be calculated from Lagrange's equations of motion without any special difficulty.

I.C Hamilton's Mechanics - Introduction to Nosé's Mechanics

William Rowan Hamilton (1805-1865) was educated in his native Ireland and joined the faculty at Dublin in 1827. His analyses of optical phenomena led to mathematical tools fundamental to the treatment of both classical and quantum mechanical systems.

In Hamilton's mechanics, just as in Lagrange's, (generalized) coordinates q can be used to simplify the treatment of (holonomically) constrained systems. In Hamilton's mechanics the conjugate momenta p (not velocities) and coordinates q appear in the Hamiltonian $H(q,p)$ on an equal and symmetric footing. Hamilton's equations of motion also differ from Newton's and Lagrange's in another way. They are *first*-order in time rather than second, giving $\dot{q}$ and $\dot{p}$ in terms of the coordinates and momenta:

$$\dot{q} = +\partial H/\partial p; \quad \dot{p} = -\partial H/\partial q.$$

Hamilton's mechanics is basic to treating quantum systems. The Schrödinger equation of quantum mechanics, which describes not only stationary states but also the time-development of a quantum system's behavior, is based upon the existence of a Hamiltonian description of that system. Like Newton's and Lagrange's equations, Hamilton's equations of motion are time-reversible. But the coordinates and momenta behave differently when time is reversed. In the reversed motion each coordinate q is unchanged in value and with successive values traced out in reversed order along the reversed trajectory. On the other hand each momentum p has to be replaced by $-p$.

In any of the three kinds of mechanics we have mentioned so far the underlying potential function, Φ, L, or H, does not depend upon the direction of time. The Hamiltonian H is usually the total energy, $K + \Phi$, expressed as a function of the q and p. Generally, it can be constructed from the Lagrangian $L(q, \dot{q})$ through the equations

$$p = \partial L(q, \dot{q})/\partial \dot{q};$$

$$H(q,p) = (\dot{q} \cdot p) - L.$$

Thus, in the case of the two-dimensional triatomic molecule considered above, the Hamiltonian has the form

$$H(\alpha, \beta, p_\alpha, p_\beta) = \left[(p_\alpha/\cos\alpha)^2 + (p_\beta/\sin\alpha)^2\right]/4m.$$

Gibbs' statistical mechanics gives the probability of finding a classical Hamiltonian system at temperature T in the phase-space region q, p within dq and dp in terms of the system's Hamiltonian function, $H(q, p)$. The corresponding "canonical-ensemble" probability density is proportional to the "Boltzmann factor" $e^{-H(q,p)/(kT)}$. This result allows us to calculate the probability distribution for the angle α defined above in the triatomic molecule problem. If we perform a canonical-ensemble average, integrating first over β (which provides only a factor of 2π), and then over the momenta, p_α and p_β, we find that the probability density $P(\alpha)$ is proportional to $\sin(\alpha)\cos(\alpha) = (1/2)\sin(2\alpha)$. This probability density has its maximum value for the right-angled configuration. On the other hand, if we avoided Lagrangian mechanics and instead forced the two distances r_{01} and r_{02} to lie close to 1 by linking the pairs of particles 01 and 02 with very stiff Hooke's-Law springs, there would be *no* coupling between the coordinates and momenta. The Hooke's-Law Hamiltonian in this case, for springs of unit rest length, would be

$$H = \left[(p_1^2 + p_2^2)/(2m) \right] + (\kappa/2) \left[(r_{01} - 1)^2 + (r_{02} - 1)^2 \right],$$

and a canonical average would give a *constant* probability density for the angle α. In **Figure 11** the rigid-constraint and the Hooke's-Law probability densities $P(2\alpha)$ are compared.

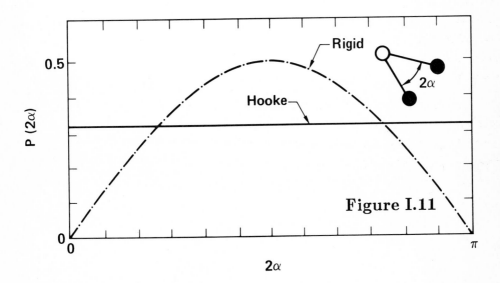

This example, which is relatively "well-known" (to the experts), illustrates that classical mechanics contains many surprises. Simply stating that we would like to consider the dynamics of a molecule with "fixed" bondlengths or fixed angles does not constitute a well-posed problem. We must specify the details of how such constraints are to be imposed.

Hamilton's equations of motion differ from Newton's in that the coordinates and momenta are *independent* variables. No longer is p necessarily equal to $m\dot{q}$. A very recent modification and extension of Hamiltonian mechanics, due to Shūichi Nosé, can be thought of as scaling either time or mass, in order to satisfy desirable constraints. In Nosé's mechanics, to be discussed in Section E of Chapter I, the time derivative of q is generally quite different from Hamilton's p/m. Let us briefly illustrate that difference by considering a harmonic oscillator with mass m and force constant κ. We first consider a Hamiltonian oscillator, for which p *is* identical to $m\dot{q}$, and next a Nosé one-dimensional oscillator, for which it is not.

In the Hamilton case the Hamiltonian is

$$H = \left[(p^2/m) + \kappa q^2\right]/2,$$

from which the familiar equations of motion follow:

$$\dot{q} = p/m; \quad \dot{p} = -\kappa q;$$

or, to emphasize the symmetry between the coordinates and momenta,

$$\ddot{q} = -\omega^2 q; \quad \ddot{p} = -\omega^2 p.$$

These Hamilton's equations, like Newton's, are "reversible". **Figures 12 and 13** illustrate a typical oscillator trajectory (with the force constant κ and mass m set equal to unity) $q = \sin(t)$; $p = \dot{q} = \cos(t)$, as well as the reversed trajectories. In the q,p,t representation of **Figure 12** reversal would correspond to an inversion through the q axis, from the point (q,p,t) to the corresponding point $(q,-p,-t)$. In the q,p representation of **Figure 13** the reversed motion shown at the bottom of the Figure corresponds to a jump, at fixed q, from p to $-p$.

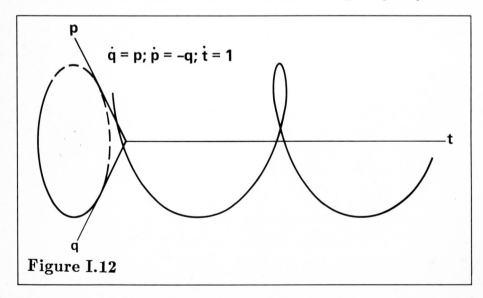

$$\dot{q} = p; \ \dot{p} = -q; \ \dot{t} = 1$$

Figure I.12

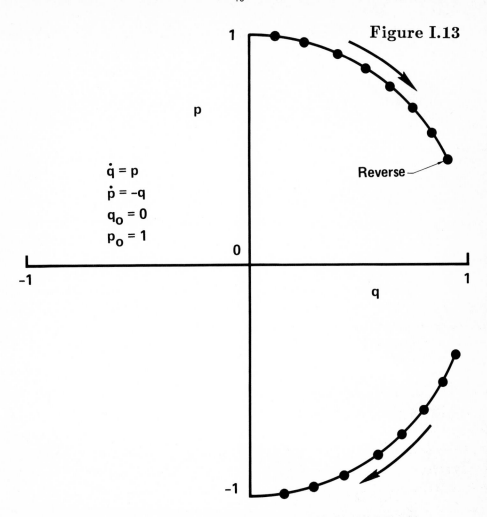

Figure I.13

$$\dot{q} = p$$
$$\dot{p} = -q$$
$$q_0 = 0$$
$$p_0 = 1$$

Reverse

An oscillator can also be described by a more complicated *constant-temperature* version of Hamiltonian mechanics which was recently invented, or discovered, by Shūichi Nosé. This important development merits its own section, Section E of Chapter I, but here we consider briefly the harmonic-oscillator example to indicate the relation between Hamilton's and Nosé's mechanics. Very little work has so far been carried out with Nosé's mechanics, either in the classical or the quantum case. But this is so only because it is new. For a classical one-dimensional oscillator Nosé's temperature-dependent Hamiltonian is

$$H_{Nos\acute{e}} = \left[p^2/(2ms^2)\right] + \left(\kappa q^2/2\right) + kT\ln s + \left[p_s^2/(2Q)\right].$$

There is a *new variable s*, which is *dimensionless*, with a *new "conjugate momentum"* p_s, which has units of *action*, in addition to the usual oscillator coordinate q and momentum p. T is the temperature and k is Boltzmann's constant. Q is a parameter with units of mass × area. A polar-coordinate form of the Hamiltonian is more useful in the quantum-mechanical case:

$$H_{Nos\acute{e}} = \left[(p_x^2 + p_y^2)/(2m)\right] + \left[\kappa Q\theta^2/(2m)\right] + (kT/2)\ln(mr^2/Q).$$

The dimensionless variable s, which becomes r in the polar representation, can be thought of as scaling the mass or scaling the time. Nosé prefers the latter interpretation. If we interpret the variable r in the polar-coordinate Hamiltonian as a length, and θ as an angle, the conjugate momenta become the conventional radial and angular momenta. Q, the parameter determining the time-dependence of the temperature (kinetic energy) fluctuations, still has units of mass times area in this form.

It is straightforward to differentiate Nosé's oscillator Hamiltonian in order to write down the equations of motion. There are four of these, one each for $q, p, s,$ and p_s:

$$\dot{q} = p/(ms^2); \quad \dot{p} = -\kappa q; \quad \dot{s} = p_s/Q; \quad \dot{p}_s = \left[p^2/(ms^3)\right] - (kT/s).$$

There is no convenient way to rewrite these first-order equations as second-order equations, so the Verlet algorithm cannot be used to solve them. Two typical (Runge-Kutta) solutions, projected onto qp space and followed for many oscillator vibration periods, are shown as the two upper trajectories in **Figure 14**. These projected q, p trajectories are not affected by time scaling.

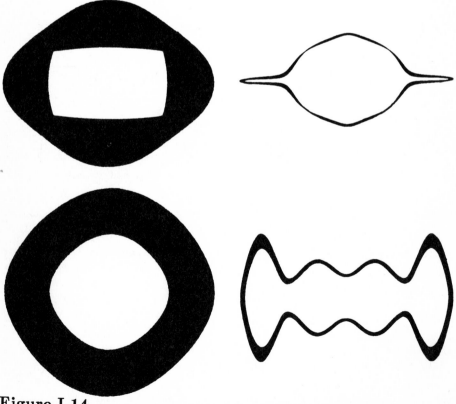

Figure I.14

But the $q, \dot{q}$ trajectories *are* changed, qualitatively, by scaling. Suppose that we pursue Nosé's time scaling idea, introducing a new time t_{new} related to the old time by the equation

$$dt_{new} = dt_{old}/s.$$

The effect of this time scaling on the four equations of motion just given is to multiply each time derivative by s. Thus, with the dots now indicating derivatives with respect to the *new* time, t_{new}, we have,

$$\dot{q} = p/(ms); \quad \dot{p} = -\kappa q s; \quad \dot{s} = sp_s/Q; \quad \dot{p}_s = \left[p^2/(ms^2)\right] - kT.$$

In **Figure 14** the $q, \dot{q}$ trajectories from these equations are shown just below the corresponding q, p trajectories *after* time scaling has been introduced. In **Figure 14** the initial values of $q, p, s,$ and p_s are respectively 1, 1, 1, and 0. The lefthand trajectory corresponds to $Q = 1$ while the righthand trajectory corresponds to $Q = 0.1$. Nosé's generalization of Hamiltonian mechanics changes the connection between the momentum p and the time derivative of the coordinate q. Because the time scale variable s is relatively small near the turning points the magnitudes of the scaled velocities $\dot{q}$ are much greater than those of the p/m. Because the trajectories are not periodic a two-dimensional region in the space is gradually filled in.

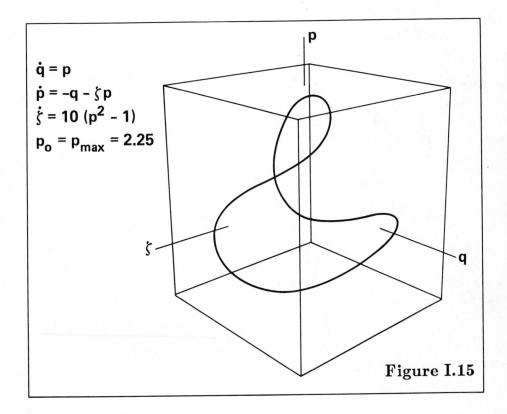

$$\dot{q} = p$$
$$\dot{p} = -q - \zeta p$$
$$\dot{\zeta} = 10\,(p^2 - 1)$$
$$p_o = p_{max} = 2.25$$

Figure I.15

Figure 15 illustrates a *periodic* solution of the scaled Nosé oscillator equations. *Any* of the three combinations of p, m, and s — $[p/(ms^2)]$, $[p/(ms)]$, or $[p/m]$ — can correspond to "momentum". (Remember that s is dimensionless.) The consequences of the three choices are quite different because s varies over such a wide range of values. (The microcanonical-ensemble fluctuations of $1/s$ diverge!) The variables q, p, s, and p_s are interrelated in the same way by either set of equations. But the trajectories are traced out at different rates in the two cases. The equations shown in the **Figure** are written in terms of the new variables, corresponding to the choice $\dot{q}_{new} = p_{old}/(ms) = p_{new}/m$, with m, κ, and kT all set equal to unity, and with Q set equal to 0.10. The variable ς in the **Figure** corresponds to the momentum p_s/Q.

Despite the close connection between Hamilton's and Nosé's mechanics, as illustrated here for the oscillator, it is possible to develop another even more useful version of Nosé's mechanics *which is entirely distinct from Hamilton's*. We will return to that subject in Section E of this Chapter. The present version of Hamilton-Nosé mechanics, like the usual Hamilton's, Lagrange's, or Newton's mechanics, uses a potential function (here H, rather than L, or Φ) to generate future behavior. The main difference between the various approaches is computational. The numerical techniques appropriate for second-order differential equations cannot always be applied to first-order equations.

I.D Gauss' Mechanics and the Principle of Least Constraint

There is a mechanics, founded on Gauss' Principle of Least Constraint, which is still more general than Newton's or Lagrange's or Hamilton's mechanics. These other three forms of mechanics can all be derived from Gauss' Principle. This Principle is particularly useful in describing the motion of constrained systems. A "rigid" diatomic molecule is probably the simplest example. In such a molecule the two atoms are constrained to remain a fixed distance apart.

Gauss' Principle can readily be applied to the triatomic molecule just treated with Lagrange's mechanics in Section B. This typical *holonomic* application presents no difficulties. Gauss' mechanics can just as easily be applied to *non*holonomic constraints, with the velocities entering in an essential way. If the constraints are either (i) holonomic or (ii) nonholonomic, but only *linear* in the velocities, then Gaussian mechanics predicts nothing new. That is, the motion predicted by Gauss would be the same as that predicted by Newton or Lagrange or Hamilton. On the other hand, for general nonlinear nonholonomic constraints, Gauss makes new predictions while Newton, Lagrange, and Hamilton are silent.

There are very few solved problems in the mechanics literature involving *nonlinear* non-holonomic constraints. We illustrate one of these museum pieces here, probably the best known, with a quadratic nonholonomic constraint:

$$\dot{x}^2 = \dot{y}^2 + \dot{z}^2.$$

This textbook example arises in the approximate treatment of the motion of "Appell's cart". The cart is shown in **Figure 16**. The knife-edged front wheel can rotate only in the plane parallel to the cart body. The motion is driven by a weight mounted on a pulley at the rear of the cart.

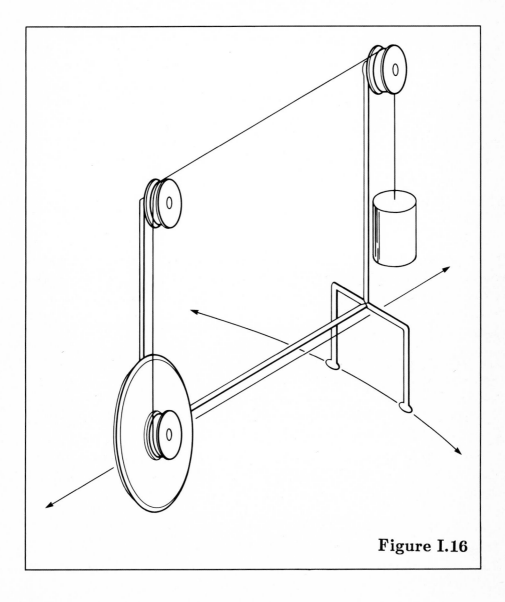

Figure I.16

The twin skids at the rear of the cart can slide, without friction, both back and forth and from side to side. The trajectories of Appell's cart are interesting. In addition to the oscillatory back-and-forth motion, the cart can spin about the bottom of the knife-edged wheel. This combination of back-and-forth with spinning motions can trace out patterns resembling the arrangements of petals in flowers.

Once we consider thermodynamic and hydrodynamic many-body systems, many nonlinear nonholonomic constraints become possible. For example, energy, stress, and heat flux are all nonlinear functions of the particle velocities. In constraining such variables Gauss' principle is useful in a unique way, inaccessible to Newton's, Lagrange's, and Hamilton's mechanics.

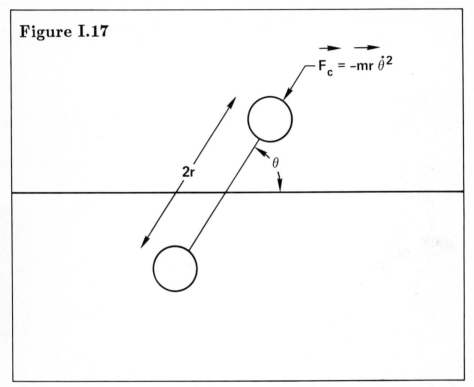

Figure I.17

What is Gauss' Principle? Gauss stated that the constraint forces F_c required to impose any constraint should be as small as possible in a least-squares sense:

$$m\ddot{r} = F + F_c;$$

$$\sum [F_c^2/(2m)] \quad \text{minimized.}$$

This is Gauss' *Principle of Least Constraint*. Let us illustrate the Principle by applying it to the simple two-dimensional rigid diatomic molecule shown in **Figure 17**. In this case Gauss' Principle produces a radial acceleration counteracting the centrifugal forces, without any tangential component. We can find the acceleration by considering two restrictions on the variation of the constraint force F_c. First, we formulate the constraint:

$$r^2/2 = \text{constant}$$

in terms of the acceleration $\ddot{r}$. To do this we calculate the second time derivative of the constraint. The result is

$$\dot{x}^2 + \dot{y}^2 + x\ddot{x} + y\ddot{y} = \dot{x}^2 + \dot{y}^2 + x\left(F_c^x/m\right) + y\left(F_c^y/m\right) = 0.$$

If we then consider a small variation of the constraint force, the variation must obey the restriction:

$$x\,\delta\left(F_c^x/m\right) + y\,\delta\left(F_c^y/m\right) = 0.$$

On the other hand, Gauss' Principle also restricts variations in the constraint force. The variational condition, $\left[(F_c)^2/(2m)\right]$ minimum, implies that the dot product of the constraint force and its variation must vanish:

$$\left(F_c^x\,\delta F_c^x + F_c^y\,\delta F_c^y\right)/m = 0.$$

Combining the two restrictions on δF_c, using a Lagrange multiplier λ chosen such that $F_c^x + \lambda x = 0$ implies that $F_c^y + \lambda y = 0$ as well, so that the second time derivative of the original constraint equation becomes

$$\dot{x}^2 + \dot{y}^2 + x\left(F_c^x/m\right) + y\left(F_c^y/m\right) = \dot{x}^2 + \dot{y}^2 - (\lambda/m)(x^2 + y^2) = 0.$$

Thus we find the value of the Lagrange multiplier, $\lambda = m\dot{r}^2/r^2$, and the corresponding constraint force, directly from the constraint equation.

This illustration of Gauss' Principle is a typical textbook one in which the energy $2m(r\dot{\theta})^2/2$ is conserved because the forces of constraint do no work. But Gauss' Principle can also be applied to more-complicated work-performing constraints involving collective thermodynamic and hydrodynamic variables. No fundamental rules or variational principles independent of Gauss' are available for such constraints. For instance it is *not* true that the work performed by the constraints should be a minimum. Constraint forces leading toward the ultimate potential minimum, the formation of a perfect crystal, would perform less work against the potential energy Φ than would any others.

Gauss' Principle is particularly useful in simulating steady nonequilibrium flows. Such flows require special methods to compensate for the natural dissipation of work into heat. The heat should be extracted to avoid the complicated description and analysis of a continually changing thermodynamic state.

For simplicity it is convenient to use Gauss' Principle to remove the heat in such a way that the nonequilibrium state is a "steady" one. By steady we mean that the driving force or the resulting flux, as well as two thermodynamic state variables (energy and density, or temperature and stress, for instance) are held constant.

To illustrate the application of Gauss' Principle to the problem of maintaining temperature constant, consider a D-dimensional N-body system. In such systems the kinetic energy typically fluctuates on the same time scale as do the particle velocities. These fluctuations correspond to fluctuations in temperature. The temperature of a many-body system can be most simply defined in terms of the kinetic energy:

$$DNkT/2 = \sum(m\dot{r}^2/2).$$

Thus, fixing the kinetic energy corresponds to fixing a dynamical estimate of the thermodynamic temperature. The derivative of the isothermal constraint just given can be written

$$\sum m\dot{r} \cdot \ddot{r} = \sum m\dot{r} \cdot \left[(F + F_c + \delta F_c)/m\right] = \sum\left[m\dot{r} \cdot (\delta F_c/m)\right] = 0.$$

Combining this with the variational form of Gauss' Principle, $\sum(F_c \cdot \delta F_c)/m = 0$, using a Lagrange multiplier ς gives

$$F_c = -\varsigma m\dot{r};$$

$$m\ddot{r} = F(r) + F_c(r, \dot{r}) = F - \varsigma m\dot{r},$$

with

$$\varsigma = \sum(F \cdot \dot{r})/\sum(m\dot{r}^2).$$

These second-order differential equations are the "Gaussian isothermal" equations of motion. Like Newton's, these equations are time reversible with the friction coefficient ς changing sign in the reversed trajectory. The friction coefficient enforces the constraint of constant temperature. Like the temperature-dependent Hamiltonian of Nosé, this is an example of something new in mechanics. Many more examples are given in Chapter IV, **Nonequilibrium Equations of Motion**. In the next section we join Nosé in straying even farther from Newton.

I.E Nosé's Mechanics - Temperature and Pressure Constraints

The classical approaches of Newton, Lagrange, and Hamilton all generate trajectories $r(t)$, $q(t)$, or $q(t)$, $p(t)$ along which energy is conserved. With periodic boundaries, linear momentum is also conserved. The ensemble approach of Gibbs' equilibrium statistical mechanics is more flexible. Energy and volume can be fixed, as in the microcanonical ensemble. Alternatively, temperature and pressure can be used as independent variables, allowing the energy and volume to fluctuate. But in these cases temperature and pressure characterize an ensemble of systems and not the individual members. It may be necessary to average over many states in an equilibrium statistical ensemble in order to estimate the corresponding temperature or pressure.

On the other hand, any treatment of nonequilibrium problems would be useless if it lacked a method for following stress, heat flux, and temperature as instantaneous functions of time in individual systems. In real laboratory experiments such variables are monitored by measurement, or inference, even under nonequilibrium conditions. In computer experiments one would also expect to measure time-dependent currents and to find that these currents vary in a reproducible way. This expectation has been abundantly justified in a variety of simulations. Thus, molecular dynamics simulations have instilled confidence that pressure and temperature can be usefully defined as instantaneous mechanical phase functions in individual systems, rather than as nebulous ensemble properties which can only be determined by exhaustive sampling.

Here we hold to the usual, and useful, viewpoint that the instantaneous temperature is defined by the mean-squared velocity, relative to the local stream velocity

$$(DNkT/2) = \sum (m/2)(v - v_o)^2.$$

With this point of view Gauss' dynamics generates a canonical ensemble of configurations, in which the probability of any configuration with potential energy Φ is proportional to $e^{-\Phi/(kT)}$, where T is the temperature just defined.

This canonical-ensemble form for the phase-space distribution can be derived from the extension of Liouville's Theorem appropriate to non-Newtonian systems described by dynamical equations such as Gauss'. In the system's phase space q, p the product of the probability density $f(q, p)$ and the differential hypervolume $dq\, dp$ is conserved by any set of equations of motion which neither creates nor destroys systems. The probability density itself flows through the phase space as a *compressible* fluid because Gaussian mechanics allows the phase-space dilatational strain-rate, $\sum [(\partial \dot{q}/\partial q) + (\partial \dot{p}/\partial p)]$, to be nonzero. Thus the probability density $f(q, p)$ obeys a generalized "continuity" ("mass" conservation) equation,

$$(\partial f/\partial t) + \sum [\partial (f\dot{q})/\partial q] + \sum [\partial (f\dot{p})/\partial p] = 0.$$

For clarity, it is worthwhile to illustrate this conservation relation with a specific example, the damped one-dimensional harmonic oscillator. For convenience we choose the mass and the force constant both equal to unity. The equation of motion is

$$\ddot{x} + \epsilon\dot{x} + x = 0,$$

where $\epsilon = 2$ corresponds to the critically-damped case. For the critically-damped oscillator

$$x = x_o\, e^{-t}$$

is a solution of the equation of motion. This critically-damped case is compared with undamped, underdamped, and overdamped trajectories in **Figure 18** with initial condition $x = \dot{x} = 1$. In the critically-damped case the phase space density follows the differential equation

$$\dot{f} = -f\,(\partial\dot{p}/\partial p) = 2f.$$

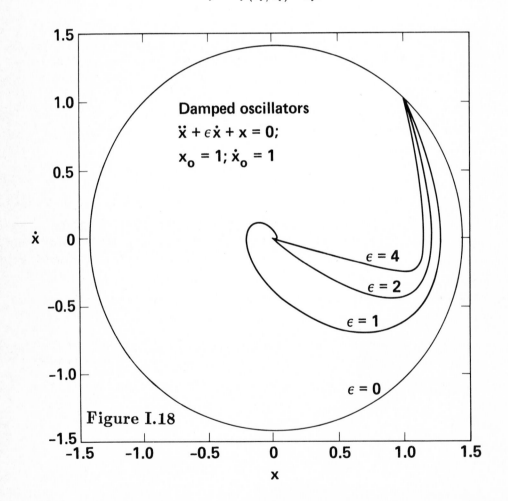

Figure I.18

Thus the critically-damped probability density *increases* exponentially in time, as the phase points near the origin become increasingly likely. The volume element $dxdp$ likewise *decreases* exponentially with time, because the product, $f \times dxdp$, is constant. Such a phase-space volume corresponds to an ensemble of critically-damped oscillators, originally distributed uniformly throughout the volume $dxdp$. As time goes on, the ensemble will occupy a phase-space volume which *decreases* exponentially in time. Because probability is conserved, the product, $f(x,p)dxdp$, is a constant of the equations of motion, namely the number of systems in the ensemble.

The compressible flow of phase-space probability leads to interesting consequences for many-body systems. Consider again our many-body Gaussian isothermal example, in which the temperature is kept fixed by a friction coefficient ς which varies with time. The probability density $f(q,p)$ in the many-body phase space varies with time because the $\dot{p}$ are explicit functions of the p. The derivative of the probability density function, following the motion, is

$$\dot{f} = (\partial f/\partial t) + \sum \left[\dot{q}\left(\partial f/\partial q\right) + \dot{p}\left(\partial f/\partial p\right) \right] =$$

$$-f \sum \left[(\partial \dot{q}/\partial q) + (\partial \dot{p}/\partial p) \right] = +3N\varsigma f = -\left[\dot{\Phi}/(kT) \right] f.$$

Thus the logarithm of the probability density rises and falls, as a function of time, in phase with the fluctuations in the potential energy. It is evident that a steady solution of this differential equation is the "isokinetic" canonical distribution

$$f \propto \delta(K - K_o)\, e^{-\Phi/(kT)},$$

where kT is $2K_o/DN$ for an N-body D-dimensional system and where the delta function guarantees that K is equal to K_o. It makes no sense to apply Gauss' constant-temperature dynamics to a single one-dimensional harmonic oscillator. Such an oscillator would have no turning point and the coordinate would diverge linearly with time. The simplest oscillator problem which *can* be treated with Gaussian constant-temperature dynamics is a *two*-dimensional harmonic oscillator. That case leads to coordinate-space trajectories resembling the floral patterns generated by Appel's cart.

Shūichi Nosé discovered a different dynamics which generates the *complete* canonical distribution, not just the configurational part. In Nosé dynamics fluctuations in the kinetic energy K are included. His equation of motion is exactly the same as the isothermal example just given using Gauss' mechanics:

$$m\ddot{r} = F - \varsigma m\dot{r},$$

but the friction coefficient ς is not solely a function of the current phase variables. Instead it depends on the *time integral* of the difference, $K - K_o$, between the kinetic energy K and its desired value K_o:

$$\varsigma = (1/Q) \int_0^t \left[\sum (p^2/m) - DNkT \right] ds.$$

The lower limit of the integral must be a fixed, but arbitrary, time—here chosen equal to 0.

Other versions of Nosé's equations can be based on higher velocity moments. For instance the equations of motion for N one-dimensional particles

$$\dot{q} = p/m; \quad \dot{p} = F - \varsigma' \left(p^3/mkT \right); \quad \dot{\varsigma}' = (kT/Q) \sum \left[\left(p^2/mkT \right)^2 - 3 \left(p^2/mkT \right) \right].$$

have also the same canonical-ensemble steady distribution in the phase space.

Nosé's mechanics is reversible in time, as are also the isoenergetic mechanics of Newton, Lagrange, and Hamilton, and the isokinetic mechanics of Gauss. Just as in the Gaussian case, in the reversed Nosé motion not only is the momentum p replaced by $-p$; but also Nosé's friction coefficient ς is replaced by $-\varsigma$. See **Figure 19** for a periodic solution of Nosé's oscillator equations of motion. Reversing the time would correspond, in **Figure 19**, to inverting the trajectory through the q axis so that the point (q, p, z) becomes $(q, -p, -z)$ in the reversed trajectory.

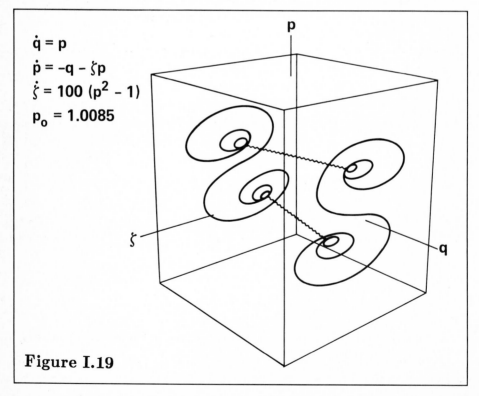

$\dot{q} = p$

$\dot{p} = -q - \varsigma p$

$\dot{\varsigma} = 100\ (p^2 - 1)$

$p_0 = 1.0085$

Figure I.19

Pedagogical derivations for Nosé's dynamics can be developed in two different ways, (i) by starting with a temperature-dependent Hamiltonian in which the variable s scales the time or the mass, or (ii) by requiring that the equation of motion generate the canonical distribution including a Gaussian distribution in the friction coefficient ς. Nosé's papers follow the former approach. Because the latter approach is not only constructive and simple, but also can be generalized to other forms of the equations of motion, we illustrate it here. We begin by considering the probability density $f(q, p, \varsigma)$ in an extended phase space which includes ς as well as all pairs of phase variables q and p. This density f satisfies the conservation of probability

$$(\partial f/\partial t) + \sum [\partial(\dot{q}f)/\partial q] + \sum [\partial(\dot{p}f)/\partial p] + [\partial(\dot{\varsigma}f)/\partial \varsigma] = 0.$$

This general continuity equation is called the "Liouville equation" in the special case that it can be simplified, using the Hamiltonian equations of motion, to the form $df/dt = 0$.

If the distribution has the form

$$f \propto e^{-Q\varsigma^2/(2kT)} f_{equilibrium},$$

and the equations of motion have a friction-coefficient form:

$$\dot{q} = p/m; \quad \dot{p} = F - \varsigma p; \quad \dot{\varsigma} = \dot{\varsigma}(q, p, Q),$$

we can calculate each of the four contributions to the phase-space flow equations. These contributions are:

$$(\partial f/\partial t) = 0;$$

$$\sum \partial(\dot{q}f)/\partial q = \sum F (p/m)(f/kT);$$

$$\sum \partial(f\dot{p})/\partial p = \sum [-pF + \varsigma (p^2 - \langle p^2 \rangle)] [f/(mkT)];$$

$$\partial(f\dot{\varsigma})/\partial \varsigma = \dot{\varsigma} (\partial f/\partial \varsigma) = -Q\varsigma (f/kT)\dot{\varsigma},$$

where we have assumed that $\dot{\varsigma}$ depends only on the phase variables, and not on ς. The conservation of probability requires that the four contributions listed above sum to zero. This requirement in turn implies that $\dot{\varsigma}$ must satisfy the equation

$$\dot{\varsigma} = \sum [p^2 - \langle p^2 \rangle]/(mQ) = 2(K - K_o)/Q.$$

If we were to use a *cubic* frictional force, $-\varsigma' p^3/(mkT)$ rather than $-\varsigma p$, we could alternatively fix the value of the *fourth* moment of the velocity distribution relative to the *second* moment, and again recover the canonical distribution. These same ideas can be generalized to include "constant-pressure" or "constant-stress-tensor" ensembles. The result is a *strain rate* (dilatational for constant-pressure, and including shear for constant-stress) which obeys a first-order relaxation equation similar to that just written for the kinetic energy.

Before leaving Nosé's mechanics, a technical point should be made. It has only been shown that the equations are *consistent* with the canonical distribution. That is, a canonical distribution in the phase space is *preserved* by Nosé's equations of motion. Whether or not the equations will *generate* such a distribution from almost all initial conditions is a subtle question. For a collisionless gas, they will not. Consider for instance a Nosé ideal gas with kinetic energy $K = \sum p^2/(2m)$. Nosé's equation of motion gives

$$\dot{K} = \sum (\dot{p}p/m) = \sum (-\varsigma p^2/m) = -2\varsigma K;$$

$$\dot{\varsigma} = -\left[(d/dt)^2 \ln(K/K_o)\right]/2 = 2(K - K_o)/Q,$$

where K_o is the specified time-averaged value of K.

If we introduce a new variable X, the logarithm of the kinetic energy ratio, K/K_o, then X oscillates in a nonlinear Toda potential:

$$\ddot{X} = -(4K_o/Q)\left[e^X - e^{X_o}\right] = -d(\Phi/m)_{effective}/dX.$$

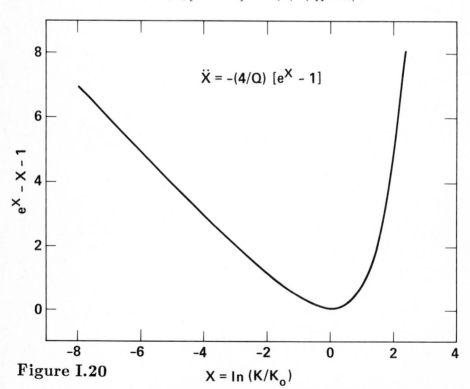

$$\ddot{X} = -(4/Q)\,[e^X - 1]$$

Figure I.20

$X = \ln (K/K_o)$

This Toda potential is sketched in **Figure 20**. Both the friction coefficient and the kinetic energy itself oscillate (with harmonic oscillations if K is close to K_o). These periodic oscillations show that a collisionless gas will never actually reach a canonical distribution under the influence of Nosé's mechanics. On the other hand, numerical work indicates that the Lyapunov instabilities associated with collisions are ordinarily sufficient to induce canonical behavior.

In general, if the parameter Q, which is proportional to the square of the heat-bath response time, is made small enough, Nosé dynamics becomes indistinguishable from Gaussian isokinetic dynamics and *never* reaches a true canonical distribution in momentum space. If Q is large the dynamics approaches Newtonian dynamics with $\langle p^2/m \rangle = DkT$. The approach to the Gaussian and Newtonian limits for the one-dimensional oscillator is indicated in **Figure 21**. The truly Gaussian case is singular, with turning points at $\pm\infty$. In **Figure 21**, with $Q = 0.01$, there are turning points just beyond $q = \pm 4$, but the limiting behavior for $Q \sim 0$ is easy to visualize. The inner Nosé trajectory shows the same solution in Nosé's original variables, without time scaling. The one-dimensional Newtonian ellipse with $\langle p^2/m \rangle = kT$ is also shown.

From **Figure 21**, it can be seen that a single harmonic Nosé oscillator doesn't achieve the canonical distribution, for *any* value of the parameter Q. And the situation is no better if cubic frictional forces, enforcing the fourth moment, are added. Thus the indications are that systems cannot be *too* simple if they are to show the phase-space mixing properties necessary to establish the canonical distribution.

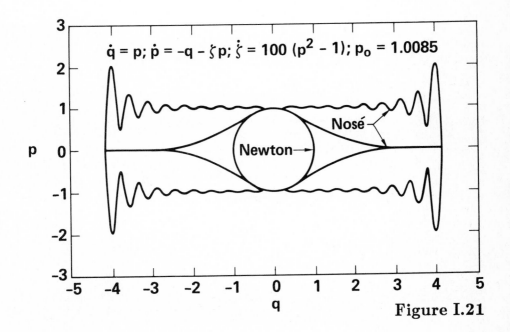

$$\dot{q} = p; \quad \dot{p} = -q - \zeta p; \quad \dot{\zeta} = 100\,(p^2 - 1); \quad p_o = 1.0085$$

Nosé

Newton→

Figure I.21

At present it isn't possible to *prove* that a particular system has the necessary mixing property. Computation can provide clues, but ultimately the problem of proof is a theoretical one. The finite precision of any computer simulation guarantees that the long-time solution of any interesting trajectory will be dominated by numerical errors. On the other hand, it *appears*, from numerical work, that a two-dimensional two-body problem *can* already exhibit canonical behavior. The precise degree of complexity required for Nosé dynamics to produce canonical behavior is not yet firmly established, even numerically. Evidently the value of Q is important to the convergence properties. If Q is very small, then the frictional reaction to disparities in the kinetic energy is rapid, and Gaussian isothermal dynamics results. In the opposite limit, when Q is very large, so that the response of the frictional forces is sluggish, the dynamics reduces to Newton's original form. All that can be said at present is this: "For *some* sufficiently chaotic problems, in two or more space dimensions, it seems that Nosé mechanics *does* generate the canonical distribution.

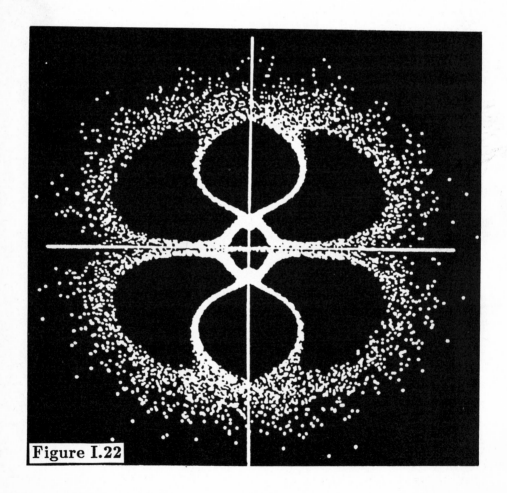

Figure I.22

The Nosé oscillator is an extremely interesting problem in its own right. For some, relatively-high-energy and small-Q initial conditions, the distribution function in the (q, p, ς) phase space has a sponge-like structure, with many holes of various shapes and sizes. **Figure 22** shows a $q = 0$ cross section (sometimes called a "Poincaré surface of section" or a "puncture plot") through a long trajectory. The occupied part of the phase space resembles a three-dimensional sponge. The holes in the sponge correspond to structures known as Kolmogorov-Arnold-Moser tori, stable regions surrounding reentrant periodic orbits. Resonances describing the coupling of pairs of these tori are responsible for the irregular sponge-like structure seen here. Such a structure is characteristic of chaotic dynamical systems.

I.F Numerical Mechanics - Fermi, Alder, Vineyard, and Rahman

With the invention of fast computers the scope of numerical integration expanded by orders of magnitude. By now, with the CRAY-2 and CRAY-XMP computers, computer capabilities have reached more than ten orders of magnitude beyond pencil-and-paper capabilities. Fast computers proliferated in the National Laboratories in the United States, first at Los Alamos, for the second World War bomb calculations, and later at Livermore, Brookhaven, and Argonne. For the first time, it became possible to solve Newton's equations for relatively complicated systems, with many degrees of freedom and for long periods of time. This capability made it possible to study numerically the irreversibility paradox that fascinated Boltzmann. This paradox can appear even in small systems such as the one-dimensional Nosé oscillator. In outline form, the reversibility paradox is as follows:

(i) From a *macroscopic* standpoint many-body systems exhibit *irreversible* behavior, as described by the second law of thermodynamics.

(ii) From a *microscopic* standpoint many-body systems are described by *reversible* equations of motion.

At Los Alamos Enrico Fermi wanted to resolve the paradox by applying the new computational tool, the computer "MANIAC", to the many-body problem. He wanted to study the approach to equilibrium of an anharmonic chain, in which the linear-force modes were coupled together by quadratic or cubic force functions of the interparticle separations.

Fermi expected to find that the entropy increase for isolated systems, predicted by the second law of thermodynamics, would follow as a consequence of Newtonian mechanics. The resulting Fermi-Pasta-Ulam calculations, mostly carried out with 16 and 32-particle anharmonic chains, showed what first appeared to be a slow approach to equilibrium, with the energy in the initially excited mode returning to its initial value less closely with each near repetition. These near repetitions occurred after dozens of vibrations of the chain, so that only a few repetitions could be accurately calculated. The early Los Alamos work did show that the number of particles

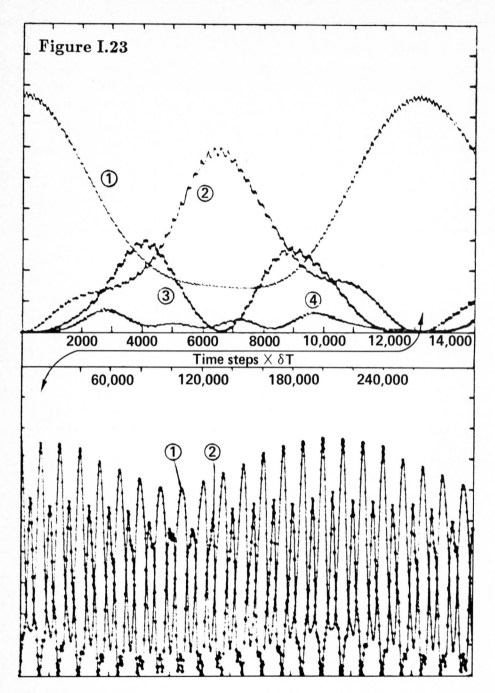

Figure I.23

was not important—17 behaved in essentially the same way as did 16 and 32—and also that the motion could be accurately reversed after several hundred time steps so as to retrace the reversed history back to the original configuration.

It is interesting that in these pioneering molecular dynamics simulations Fermi, Pasta, and Ulam did not even describe the algorithm used to integrate the equations of motion. It was assumed that the reader could work out such a scheme. The centered-difference "Verlet" scheme used by Feynman or the even better fourth-order Runge-Kutta method would be good choices.

Later calculations, carried out by Tuck and Menzel at Los Alamos, showed that the Fermi-Pasta-Ulam chains do not, in general, equilibrate and that statistical mechanics is therefore not generally valid for these oversimplified systems. See **Figure 23**. It shows a typical variation of mode energies with time. The curve labelled "1" is the amplitude of the energy of the initially-excited lowest-frequency "mode". The initial condition nearly recurs after about 13,000 time steps, as shown in the top of the **Figure**. The recurrence is even closer after the 200,000 timestep *superperiod* discovered by Tuck and Menzel. The somewhat pathological character of these systems which don't equilibrate continues to stimulate the interest of mathematicians even today. The Nosé oscillator is probably the simplest such system.

There is a reference to some unpublished *two*-dimensional many-body calculations by Fermi at Los Alamos, but the first (and very extensive) published simulations in two and three dimensions were carried out by Berni Alder and Tom Wainwright at the Lawrence Livermore "National" Laboratory (then the "Radiation" Laboratory). Alder and Wainwright wanted to see whether or not the reversible equations of motion of Newton could account for the irreversible behavior described by the Boltzmann Equation and the second law of thermodynamics.

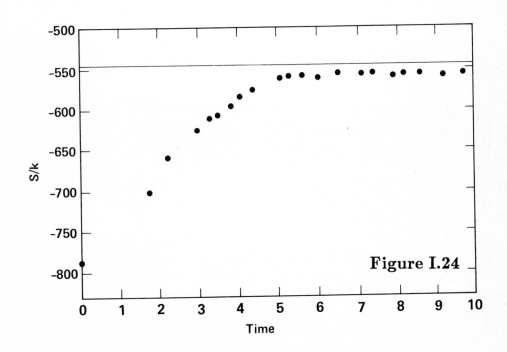

Figure I.24

Figure I.25

JOURNAL
OF
APPLIED PHYSICS

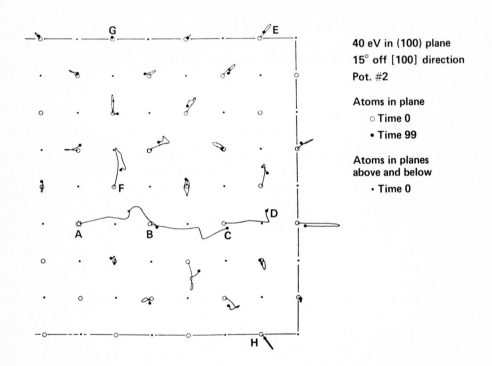

40 eV in (100) plane
15° off [100] direction
Pot. #2

Atoms in plane
○ Time 0
• Time 99

Atoms in planes
above and below
· Time 0

Alder and Wainwright studied the motion of 100 hard spheres, all with the same initial speed, but with different velocities. Their calculation of the low-density approximation to the entropy $S = -Nk \langle \ln f(p,t) \rangle$, where f is the *one*-particle probability density in momentum space, is shown in **Figure 24**. They found that the spheres established an equilibrium momentum distribution relatively rapidly, in about three collisions per particle. The resulting equilibrium thermodynamic properties also agreed with the Monte-Carlo statistical predictions generated by Alder and Wainwright's collaborators, Wood and Parker, at Los Alamos.

Alder and Wainwright also predicted the rate at which Boltzmann's low-density form for the entropy $-Nk \langle \ln f(p,t) \rangle = S(t)$, would rise to the equilibrium value according to the

Boltzmann equation. Their prediction was fully consistent with the results of the molecular dynamics simulation. At that time it was not generally recognized that the hard-sphere behavior incorporated the typical Lyapunov instability, while Fermi's anharmonic chain did not. So the early computer results were a little puzzling, with statistical mechanics and the second law of thermodynamics working at Livermore, but failing at Los Alamos.

At the Brookhaven Laboratory George Vineyard and his coworkers were interested in an application of molecular dynamics to metals damaged by radiation. This class of problems has remained important in the design of reactors. The Brookhaven calculations were fully three-dimensional, used continuous forces, and incorporated viscoelastic boundary particles. The earliest literature reference to this work seems to be the cover of the August 1959 issue of the Journal of Applied Physics. The cover, reproduced as **Figure 25**, shows the trajectories of several metal atoms in a crystal. Inside the issue there is no accompanying article, only a caption identifying the workers and a very brief description of the problems being studied.

At the Argonne Laboratory, Anees Rahman was carrying out an ambitious simulation of liquid argon, using periodic boundary conditions and 864 particles, for a long time the world's record. Rahman's calculation was the first attempt to study liquid physics with continuous potentials, focusing on structural information that could be directly compared with experiment. This emphasis followed Vineyard's in seeking to simulate particular materials. Fermi's and Alder's calculations, on the other hand, sought to elucidate mechanisms underlying general properties of simple materials.

Rahman's innovative work was honored at an Argonne Laboratory Festschrift in 1984. Alder celebrated his 60th birthday in 1985. So molecular dynamics is well on its way to becoming a "mature" field. The fact that Rahman used 864 particles, rather than a smaller number, is to some extent responsible for the relatively large number of particles that many subsequent investigators used. As we will soon point out, with several examples, both the equilibrium and the transport properties of simple systems depend only weakly on the number of particles. Because the equilibration time increases roughly as $N^{2/3}$, the diffusion time across an N-particle system, it is often more efficient to carry out a longer calculation on a smaller system than a shorter large-size calculation. A considerable amount of work has been carried out on the number-dependence of computer-experiment results. We discuss number-dependence in Section E of Chapter II.

These four sets of workers, at Argonne, Brookhaven, Livermore, and Los Alamos, set the stage for what is now called "molecular dynamics" (despite the fact that until recently most of the calculations involved monatomic interactions), an enterprise now carried on in a hundred institutions in a dozen countries, and having tremendous impact on the development of the theoretical description of classical many-body systems.

Figure I.26

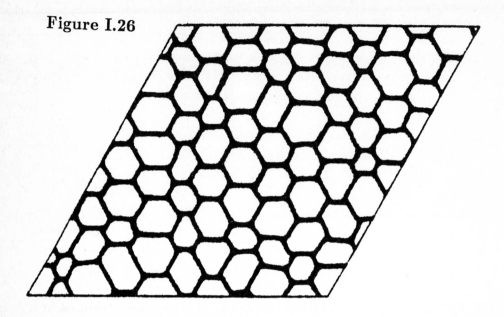

By 1985 Farid Abraham and his coworkers at IBM San Jose had carried out long simulations using 161,604 particles. **Figure 26** shows the cell structure formed by a monatomic layer of rare-gas atoms adsorbed on graphite. This was done by plotting the positions of those atoms which were out of register with the underlying graphite lattice. The calculations are unusual in that the number of atoms treated in the computer experiment is approximately the same as that observed in corresponding laboratory experiments.

Bibliography for Chapter I.

The Lyapunov instability present in the equations of motion of most interesting systems is discussed in two recent books on nonlinear dynamics, Hao Bai-Lin's *Chaos* (World Scientific, Singapore, 1984) and Heinz G. Schuster's *Deterministic Chaos* (Physik-Verlag, Weinheim, 1984).

Current papers in molecular dynamics can be found by scanning the Journal of Chemical Physics, the Journal of Statistical Physics, Physica, the Physical Review, and Physical Review Letters. The early molecular dynamics calculations are summarized in a popular article, "Molecular Motions", by B. J. Alder and T. E. Wainwright, in the Scientific American for October, 1959. This article attracted me to Livermore in 1962.

Equilibrium applications of the "Nosé thermostat" are described in two clear papers by Shūichi Nosé: "A Unified Formulation of the Constant-Temperature Molecular Dynamics Methods", Journal of Chemical Physics **81**, 511 (1984) and "A Molecular Dynamics Method for Simulations in the Canonical Ensemble", Molecular Physics **52**, 255 (1984). See also W. G. Hoover, "Canonical Dynamics: Equilibrium Phase-Space Distributions", Physical Review **A 31**, 1695 (1985) and H. A. Posch, W. G. Hoover, and F. J. Vesely, "Dynamics of the Nosé-Hoover Oscillator: Chaos, Order, and Stability", Physical Review **A 33**, 4253 (1986).

A comprehensive and pedagogical treatment of holonomic constraints can be found in a paper "Constraints", by N. G. van Kampen and J. J. Lodder, American Journal of Physics **52**, 419 (1984). Nine different forms for nonholonomic constraint forces are compared in treating a heat-flow simulation in the review by D. J. Evans and W. G. Hoover, "Flows Far From Equilibrium *via* Molecular Dynamics", Annual Review of Fluid Mechanics **18**, 243 (1986). See also Section D of Chapter IV of these notes.

The "Feynman Lectures", R. P. Feynman, R. B. Leighton, and M. Sands, *Feynman Lectures on Physics* (Addison-Wesley, Reading, Massachusetts, 1964) contain not only the finite-difference calculations mentioned in the chapter, but also an interesting chapter on Hamilton's Principle of Least Action.

For applications of Gauss' Principle see "Nonequilibrium Molecular Dynamics *via* Gauss' Principle of Least Constraint", by D. J. Evans, W. G. Hoover, B. H. Failor, B. Moran, and A. J. C. Ladd, Physical Review **A 28**, 1016 (1983). See also Section G of Chapter IV of these notes.

The Fermi-Pasta-Ulam calculation is extensively reviewed and extended in an article by J. L. Tuck and M. T. Menzel, "The Superperiod of the Nonlinear Weighted String (Fermi-Pasta-Ulam) Problem", Advances in Mathematics **9**, 399 (1972).

The 161,604-particle simulation is described in F. F. Abraham, W. E. Rudge, D. J. Auerbach, and S. W. Koch, "Molecular-Dynamics Simulations of the Incommensurate Phase of Krypton on Graphite Using More Than 100,000 Atoms", Physical Review Letters **52**, 445 (1984).

II. CONNECTING MOLECULAR DYNAMICS TO THERMODYNAMICS

II.A Instantaneous Mechanical Variables

The two main goals of microscopic molecular dynamics calculations are to simulate and to understand macroscopic behavior in microscopic terms. We wish to understand the way in which the relatively complicated microscopic many-body dynamics gives rise to the relatively simple macroscopic few-variable behavior described by phenomenological thermodynamics and hydrodynamics.

Some of the variables useful for describing macroscopic systems have obvious microscopic analogs. The macroscopic mass $\rho \, dr$ and momentum $\rho v \, dr$ in a volume element dr, for instance, correspond to simple sums of one-particle contributions. The macroscopic energy and the pressure tensor are more complicated functions. They include not only one-particle *kinetic* parts, but also two-or-more-particle *potential* contributions.

The entropy and free energy functions are even more complicated than mass, momentum, energy, and pressure. For small few-particle volume elements there is no sensible and appealing instantaneous definition of an entropy guaranteed to resemble thermodynamic entropy. But for *large* volume elements, at least sufficiently close to equilibrium, Gibbs showed that the entropy corresponds to the logarithm of the available phase-space volume. This phase-space volume can only be determined by explicitly carrying out calculations over an interval of time or by integrating over the appropriate phase space. Thus, the entropy depends upon the current state of the system in a relatively complicated way. Fluctuations likewise involve either time or phase-space averaging, and are more complicated to evaluate than mass or momentum or energy sums.

In transient nonequilibrium systems, far from equilibrium, it is not practical to define instantaneous properties in terms of constrained time or phase-space averages. This is because such systems change with time, so that the variables constraining a "phase-space-average" are not apparent. Nevertheless, no useful description of time-dependent nonequilibrium behavior is possible without *some* recipe for describing the instantaneous state of a system. Accordingly, we here consider the overall variables describing a microscopic many-body system.

Energy, volume, and the number of particles are the independent variables describing either an isolated system or a microcanonical ensemble of such systems. For any individual closed and isolated system, obeying Newton's, or Lagrange's, or Hamilton's equations of motion, these properties are constants, "constants of the motion". The only other known constant of the motion, for most interesting interparticle force laws, is the momentum. It is convenient to divide the other *non*conserved macroscopic variables needed to describe a many-body system into two categories.

The first category includes "mechanical variables". The mechanical variables of a dynamical system can be usefully defined as instantaneous functions of the sets of coordinates and velocities, r and $\dot{r}$, or the coordinates and momenta, q and p. These mechanical variables include not just mass, momentum, and energy but also the fluxes of these quantities. Provided that the interparticle interactions are pairwise-additive, the mechanical variables most useful to a hydrodynamic description can all be expressed in terms of coordinates and momenta by using simple one-body and two-body functions.

There is a second category of macroscopic variables, involving the thermodynamic entropy, which could be termed "entropic variables". These depend upon Gibbs' statistical definition of thermodynamic state, and include an entropy contribution based on a phase-space volume. The thermodynamic entropy, as well as the Gibbs and Helmholtz free energies, are examples taken from this second category. We will discuss both categories, mechanical and entropic, in turn.

Three of the fundamental mechanical variables correspond to the zeroth, first, and second moments of the microscopic velocity distribution function. The zeroth moment is proportional to the mass density ρ. The first moment is proportional to the stream velocity v. The second moment, giving the fluctuation of the microscopic velocities about the mean, is, at equilibrium, proportional to the thermodynamic temperature T. This same equilibrium temperature definition serves as a convenient and consistent nonequilibrium generalization of temperature to far-from-equilibrium states.

Temperature arises in two different ways in the microscopic statistical theory of equilibrium systems. In that theory it is usual to consider a "canonical ensemble" of similar systems weakly coupled together in such a way as to share a fixed total energy. If the probability distribution of the coupled systems over their states is expanded around the most likely distribution, temperature emerges as the derivative $(\partial E/\partial S)_V$. Here the entropy S is defined as the product of $-k$, where k is Boltzmann's constant, multiplied by the average value of the logarithm of the phase-space volume accessible to an N-body system in the volume V with an energy E. The alternative, but equivalent, Lagrange-multiplier calculation of the distribution of maximum probability, using the constraint of fixed total energy for the ensemble, produces $1/kT$ as the corresponding Lagrange multiplier.

In equilibrium thermodynamics temperature is introduced through the ideal-gas thermometer. This thermodynamic definition is more suggestive than the statistical phase-space definition because the second moment, relative to the mean, of the D-dimensional gas velocity distribution, $\langle p^2/m \rangle = DkT$, is common to *all* classical equilibrium systems, not just ideal gases, and characterizes the *whole* equilibrium distribution, because the equilibrium distribution is Gaussian. The one-dimensional equilibrium distribution is shown in **Figure 1**. The arrow, at the inflection point, indicates the momentum p_x for which p_x^2 equals mkT. The use of $\langle p_x^2 \rangle/(mk)$ to *define* a dynamic temperature is easy to defend. Given the need to do something, we follow Occam in making this, the simplest choice.

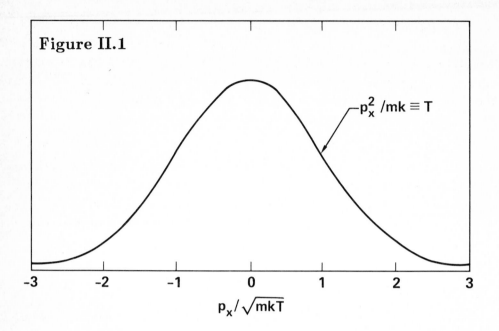

Figure II.1

$-p_x^2/mk \equiv T$

$p_x/\sqrt{mkT}$

It is an empirical result of equilibrium thermodynamics that specifying the temperature, volume, and composition is a complete fluid-state description. All other thermodynamic variables—pressure, energy, and entropy for instance—are coupled together through the empirical equilibrium "equation of state". An isolated nonequilibrium system relaxes toward such an equilibrium state during a characteristic time of the order of the time required for mass, momentum, or energy to diffuse across the system.

The number of such variables required to describe a nonequilibrium "state" sufficiently accurately has to be determined empirically. Any such state description must involve at least those variables required in equilibrium thermodynamics, plus the additional variables necessary to provide a reproducible description of the deviation from equilibrium.

In nonequilibrium systems the main requirement useful state variables must satisfy is that these be *mechanical*, depending upon the particle coordinates and momenta, rather than entropic. This is because so little progress has been made in defining and calculating entropic variables for nonequilibrium systems. Entropy is certainly more complicated to define than is temperature. A generation ago, Jaynes suggested using Gibbs' equilibrium expression for the entropy, $S = -k\langle \ln f \rangle$—where f is the N-particle phase-space probability density—even for states far from equilibrium. This idea is the "information-theory" approach to statistical mechanics. The basis, choosing as many states as possible consistent with known restrictions on the system, is undisputable but unworkable.

The mathematics involved in using this definition is relatively involved because f, from the *mathematical* standpoint, is a constant of the motion. Thus the entropy too would be constant if it were calculated exactly. But this mathematical view of the entropy leaves out the "Kolmogorov entropy", that part of the entropy production which comes from the exponentially-fast diffusive motion of f in the phase space due to Lyapunov instability. This diffusive motion persists until an equilibrium or steady distribution has been reached.

At present no simpler useful nonequilibrium prescription for entropy than Jaynes' is available. But it *is* possible that suggestions will emerge, motivated by the results of nonequilibrium computer simulations. Entropy has an apparent connection to the friction coefficient ς, which appears in Gauss' or Nosé's generalized equations of motion described in Sections D and E of Chapter I. In either case the derivative of the logarithm of the phase-space density function, $f(q,p)$, with time, following the motion, is just

$$(d/dt)\ln f = -\sum\left[(\partial\dot{q}/\partial q) + (\partial\dot{p}/\partial p)\right] = DN\varsigma.$$

Thus, if we use the Boltzmann-Gibbs connection between $-k\langle\ln f\rangle$ and entropy, the thermodynamic friction coefficient measures directly the rate of entropy production, excluding the Kolmogorov instability entropy. This relationship is potentially useful in nonequilibrium systems, but has not so far been applied to estimate nonequilibrium free energies.

In principle, equilibrium or nonequilibrium free energies could be calculated from $S = -k\langle\ln f\rangle$ if the phase-space probability density were known. In practice, too much time is required to visit, even once, each of the states of a system of any size at all, so as to find the entropy associated with that distribution. Twelve liquid argon atoms, with a triple-point entropy of $77k$, require approximately the age of the universe to travel through all of their 10^{33} (quantum) states. Even the simple one-dimensional classical Nosé oscillator requires a time of this same order, using a current CRAY computer, to reach a phase-space state lying only 11 standard deviations from the most-likely zero-energy state.

An alternative to defining the entropy through state-counting can be based on a working definition of the free energies as equilibrium values corresponding to the instantaneous energy, density, and the fluxes. But because we will have no need for such a definition in these lectures, and, because this idea has never been used, we will not elaborate on this possibility.

There *is* a special class of systems, "hard-particle" systems, for which entropy and the free energies *can* be estimated as instantaneous dynamical variables. In dense fluids or solids of hard spheres (or disks, in two dimensions) the space available for an additional sphere, a kind of "free volume" made up of holes distributed through the system, gives a direct measure of the instantaneous work required to insert another sphere:

$$\mu = (\partial A/\partial N)_{V,T}.$$

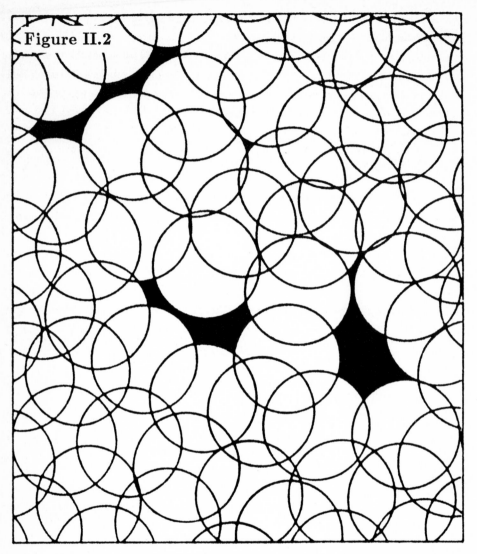

Figure II.2

Figure 2 shows a typical configuration of 48 hard disks taken from an equilibrium simulation. The density was half the close-packed density and periodic boundaries were used. The region lying outside the large "exclusion disks" is the free volume v_f available for the center of a 49th disk. The "excess" chemical potential μ^{EXCESS} (measured relative to that for an ideal gas at the same density and temperature) can be evaluated from the equilibrium average of v_f:

$$\langle v_f \rangle = V\, e^{-\mu^{EXCESS}/(kT)}.$$

Whether or not this recipe is useful in nonequilibrium systems such as far-from-equilibrium shear flows would be interesting to test. For sufficiently large volumes, free volume fluctuations can be ignored. Then the many-body state-counting problem reduces to estimating the probability of inserting one additional particle. Such an estimate involves errors of order $(\ln N)/N$.

II.B Macroscopic Dynamics

The macroscopic description of continuum systems makes use of the mass density ρ, the stream velocity v, the pressure tensor P, the energy per unit mass e, and the heat flux vector Q as fundamental variables. The three basic dynamical equations connecting the time variation, conservation, and flow of mass, momentum, and energy can be written in terms of these variables. The equations have their most transparent, compact, and elegant form when written in terms of the so-called "Lagrangian variables", using coordinates which follow the motion. These coordinates can alternatively be called Heisenberg-picture or comoving coordinates.

The "Lagrangian coordinates" have no connection with Lagrange's equations of motion. **Figure 3** illustrates the difference between comoving "Lagrangian" coordinates and inertial-frame "Eulerian" coordinates. The two-dimensional disk shown in **Figure 3** rotates counter clockwise about its center at angular velocity ω. The comoving Lagrangian coordinate system keeps pace with the rotation, rotating with this same angular velocity. In the comoving Lagrangian system the mass and energy fluxes are both zero. The momentum flux is not zero because it contains centrifugal force contributions from the frame's rotation.

The same problem can be considered in a fixed-frame coordinate system, corresponding to the Schrödinger picture of quantum mechanics. In this inertial fixed-frame coordinate system— usually called the "Eulerian frame"—the mass and energy fluxes are nonzero. In the Eulerian frame these fluxes are vectors in the θ direction, with magnitudes $\rho\omega r$ and $\rho\omega r[e + (\omega^2 r^2/2)]$.

The comoving Lagrangian derivatives which follow the motion are sometimes termed "convective" or "substantive" or "total" derivatives. Such derivatives are conventionally indicated by D/Dt, by d/dt, or by a superposed dot. For example $\dot{\rho} = d\rho/dt$ gives the change in the mass density of a volume element dr following the motion of the element.

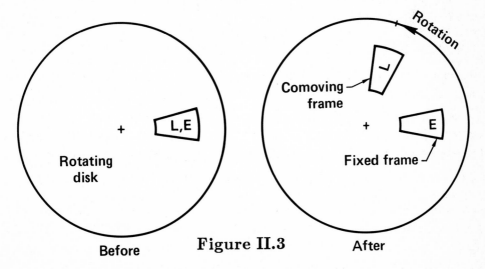

Figure II.3

Velocity vanishes in a comoving frame. It is understood therefore that the symbol v, for velocity, refers to an inertial frame. Thus v is the rate at which the velocity of a volume element (measured in an inertial frame) changes with time. Derivatives describing changes of density, velocity, and energy density, $(d\rho/dt)$, (dv/dt), and (de/dt), can be expressed in terms of the corresponding Eulerian derivatives, $(\partial\rho/\partial t)$, $(\partial v/\partial t)$, and $(\partial e/\partial t)$, by including the convective contributions proportional to the inertial-frame velocity v:

$$d/dt = \partial/\partial t + v \cdot \partial/\partial r.$$

The fundamental continuum equations describe the conservation of mass, momentum, and energy. Mass conservation is simplest. In the comoving Lagrangian coordinate system the mass of a moving volume element, $\rho\,dr$, is a constant of the motion, so that the "continuity equation",

$$(d\ln\rho/dt) = -\nabla \cdot v,$$

follows from the observation that the volume strain-rate, $d\ln V/dt$, is measured by the divergence of the velocity.

$$d\ln V/dt = (\partial v_x/\partial x) + (\partial v_y/\partial y) + (\partial v_z/\partial z).$$

The continuity equation can be derived from the inertial-frame "Eulerian" picture. In the Eulerian treatment the coordinate mesh is fixed in space, rather than embedded in the flowing materials, and the fluid or solid moves *through* the mesh. Thus, provided that the flow velocity v is a smooth function of the coordinates, the flow into a sufficiently small fixed cube $dr = dx\,dy\,dz$ is given by the flux differences across the cube:

$$\partial(\rho\,dr)/\partial t = -dx\,(\partial/\partial x)(\rho v_x\,dy\,dz) - dy\,(\partial/\partial y)(\rho v_y\,dz\,dx) - dz\,(\partial/\partial z)(\rho v_z\,dx\,dy).$$

Figure 4 shows the two-dimensional version of this flow problem, in which the element of "volume" is the area $dx dy$. The hatched regions represent mass entering and leaving the volume element during a time dt chosen sufficiently small that the vertical motion in the hatched regions can be ignored. Because the Eulerian volume element $dr = dx\,dy\,dz$ doesn't change with time, a simple differential equation results when the flux difference relation is divided by dr:

$$(\partial\rho/\partial t) = -v \cdot \nabla\rho - \rho(\nabla \cdot v).$$

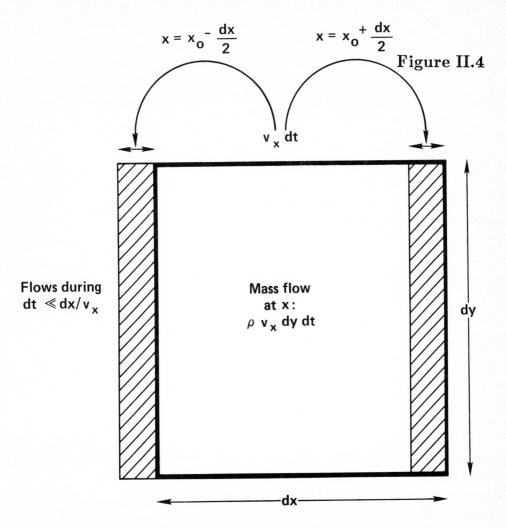

$$x = x_o - \frac{dx}{2} \qquad x = x_o + \frac{dx}{2}$$

Figure II.4

$v_x \, dt$

Flows during
$dt \ll dx/v_x$

**Mass flow
at x :**
$\rho \, v_x \, dy \, dt$

dy

dx

This equation is the Eulerian version of the continuity equation. It relates the change of density with time at a fixed location $\partial\rho/\partial t$ to the mass-density gradient and the divergence of the velocity field. Combining the two density derivatives in the Eulerian continuity equation reproduces the Lagrangian one.

The equivalence of the alternative Lagrangian and Eulerian approaches applies not only to the transport of mass, but also to the transport of momentum, energy, or any other property carried by a continuum. Both coordinate types can be useful. The Lagrange coordinates are particularly useful in numerical simulations involving large deformations, shockwaves, or interfaces separating materials, as long as the shear and rotation rates are not too large. Under such extreme conditions, the Eulerian coordinates are cumbersome and inconvenient. But problems involving *very* large deformations, or deformation rates, can lead to the tangling of hydrodynamic zones. In such a case neither choice of coordinates, Eulerian or Lagrangian, is troublefree.

Conservation of momentum in a continuum leads to the "equation of motion"

$$\rho\, dv/dt = -\nabla \cdot P,$$

where the comoving momentum flux P is the "pressure *tensor*", and includes *all viscous con-tributions*. In this Lagrangian form of the equation of motion, the comoving momentum flux is measured in a Lagrangian frame moving with an inertial-frame velocity v. Because momentum flux has two directions associated with it—one for the *direction* of the flow (x, y, or z) and one for the *type* of momentum (x, y, or z)—the tensor is a "*second*-rank" tensor, with doubly-subscripted elements. The element P_{ij} of the tensor P represents the flux, in the i direction, of j momentum. Equivalently P_{ij} is equal to the force per unit area exerted on the surrounding fluid by the ith face of an infinitesimal cube in the direction j.

Figure II.5

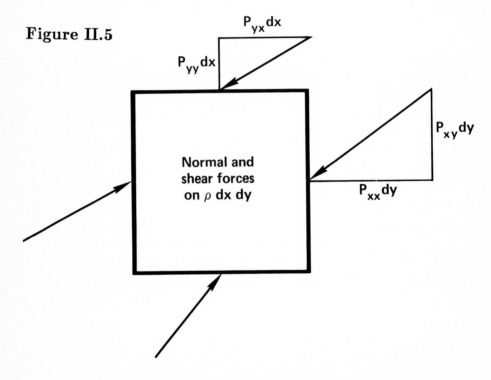

Figure 5 shows a small two-dimensional "volume" element $dx\, dy$. In a comoving frame the forces exerted on the boundary of this element by the surrounding continuum are proportional to the appropriate pressure-tensor elements and to the magnitudes of dx and dy. For a sufficiently small volume element the forces exerted on any of the four edges can be written as a superposition of perpendicular and parallel contributions as is shown in **Figure 5**. The forces exerted on the rightmost edge would be $-P_{xx}\, dy$ and $-P_{xy}\, dy$, respectively. The minus signs follow because P_{xx} and P_{xy} are fluxes of x and y momentum moving in the positive x direction, that is, *leaving* the volume element.

It might appear that the normal forces would act to accelerate linear momentum and that the shear forces would excite angular momentum, but the symmetry of the pressure tensor requires that $P_{zy} = P_{yz}$, for instance, so that the shear stresses $(P_{zz} - P_{yy})/2$ and P_{zy} act to change the *shape* of the volume element, not its rotational velocity. It is an interesting exercise to show that an imbalance of P_{zy} and P_{yz} would lead to a *divergent* angular acceleration (proportional to L^{-2} for an infinitesimal D-dimensional cube with sidelength L).

The derivation of the continuum equation of motion follows by considering the sum of the forces exerted by the surrounding continuum on a comoving cube (square in the two-dimensional case that was shown in **Figure 5**). Such a cube can be accelerated in the x direction if, for instance, the x forces on the x faces at $x - \frac{1}{2}dx$ and $x + \frac{1}{2}dx$ are not equal. In that case $\rho \, dr \, dv_x/dt$ includes $dy \, dz \, [P_{xx}(x-\frac{1}{2}dx) - P_{xx}(x+\frac{1}{2}dx)]$. In two dimensions the dz is absent and the units of momentum flux are [mass/time] rather than [mass/(length×time)]. The force difference approaches $-dr \, (\partial P_{xx}/\partial x)$ for a sufficiently small cube and a sufficiently smooth pressure tensor. Including also the x forces on the y and z faces gives the complete x component of the equation of motion:
$$(\rho \, dr) dv_x/dt = -dr \, [(\partial P_{xx}/\partial x) + (\partial P_{yx}/\partial y) + (\partial P_{zx}/\partial z)].$$
Dividing by dr then leads to the Lagrangian form of the equation of motion given previously.

Let us now consider the momentum flux P from the more-detailed atomistic viewpoint. On a microscopic basis there are two different kinds of contributions to the pressure tensor, "kinetic" and "potential". To describe the kinetic part we use the notation p to indicate the momentum carried by a particle relative to the comoving Lagrangian frame. Each Particle i in the comoving volume dr carries momentum p_i. During a small time interval dt this momentum is transported a distance $(p_i/m_i) \, dt$. This means that, on the average, a comoving plane, with area dydz and perpendicular to the x axis will intersect the moving momentum with probability $(dy \, dz/dr) \, p_x \, dt/m$ during the time interval dt. The resulting "kinetic" contribution to the momentum flux, summed over all particles within the volume element dr is

$$P_{kinetic} = \sum (pp/m)/dr.$$

An additional "potential" flow of momentum, between all pairs of particles within dr, occurs through the mechanism of the interparticle forces. Consider the pair of particles, Particle 1 and Particle 2, shown in **Figure 6**. The rate at which momentum is transported from Particle 1 to Particle 2, the force on Particle 2 due to Particle 1, is given by the pair force $F_{21} = -F_{12}$. For each such ij pair the direction of this transport is along the line of centers, parallel to $r_{ij} = r_i - r_j$.

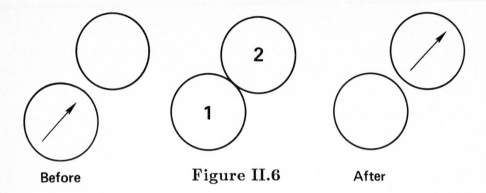

| Before | Figure II.6 | After |

The contribution of all pairs of particles to the pressure tensor is the "potential" contribution to the momentum flux:

$$P_{potential} = \sum F_{ij} r_{ij}/dr.$$

The action-at-a-distance interactions described by the interparticle forces give the microscopic pressure tensor, $P = P_{kinetic} + P_{potential}$, a nonlocal character. The details of treating pairs of particles which lie only partly in dr can lead to different definitions of $P(r)$ and hence to different macroscopic constitutive relations. This is important in problems involving surface tension, where curvilinear coordinates seem natural for describing the pressure tensor. A similar ambiguity arises if we attempt to go beyond a linear constitutive theory of nonequilibrium transport.

Conservation of energy in a continuum leads to the "energy equation"

$$\rho \, dr \, \dot{e} = dr \left[-P : \nabla v - \nabla \cdot Q \right],$$

where e is the energy per unit mass and Q is the "heat flux vector". The heat flux vector measures the conductive flow of energy, per unit time and area, in the Lagrangian comoving frame. The energy equation is a straightforward generalization of the first law of thermodynamics, which gives the energy change in a volume element as the sum of the work done on that element by its surroundings less the heat leaving the volume element through comoving conduction. The relative motion of the x faces in the x direction gives a contribution to the work per unit volume, $\left[-P_{xx} \, dy \, dz \, (\partial v_x/\partial x) \, dx \right]/dx \, dy \, dz$. The sum of the nine separate terms of this type obtained by summing contributions of $x, y,$ and z forces in the $x, y,$ and z directions are indicated by the dyadic product $-P : \nabla v$.

From the microscopic point of view the heat flux vector, like the pressure tensor, contains both kinetic-energy and potential-energy contributions. If we retain the simplest possible assumption, that the energy of interaction of two particles is shared equally by both members of the pair, then each Particle i has an energy of the form

$$E_i = \left[p_i^2/(2m_i)\right] + \sum(\phi_{ij}/2).$$

where the sum includes all Particles j with which Particle i interacts. The flow of this energy, for all particles in the volume dr, gives the kinetic contribution to the energy flux:

$$dr\,Q_{kinetic} = \sum(p_i/m_i)\,E_i.$$

The potential contribution of the interparticle forces to the flow of energy can best be visualized by considering the example pair of hard-disk particles shown in the preceding **Figure 6**, Particles 1 and 2. During a simple head-on collision of these two disks the potential energy transports the (kinetic) energy $p^2/2m$ through one particle diameter. The general expression for the flow of energy associated with this mechanism has the form

$$dr\,Q_{potential} = \sum F_{ij}r_{ij} \cdot (p_i + p_j)/(2m).$$

where the sum includes all pairs of particles in the volume element dr.

With mechanical definitions of mass density, stream velocity, energy density, pressure tensor, and heat flux, we have the tools necessary to relate the microscopic and macroscopic descriptions of matter to each other. In deriving the continuity equation, equation of motion, and energy equation, our reasoning has been macroscopic, treating materials on a continuum basis. We also indicated the way in which atomistic flow mechanisms make it possible to define the fluxes as sums of one and two atom contributions. In the next section we consider the formal structure of the pressure tensor and heat flux vector from the detailed microscopic point of view.

II.C Virial Theorem and Heat Theorem

In order to connect the microsopic pressure tensor P and heat flux vector Q to particle coordinates and momenta, we will examine the well-known virial theorem and a close relative we will call the "Heat Theorem". The usual derivation of the virial theorem of statistical mechanics begins with the canonical partition function

$$Z(N,V,T) = e^{-A/(kT)} = \int dr^{DN} \int dp^{DN} e^{-(K+\Phi)/(kT)}/N!\,h^{DN},$$

where $A = E - TS$ is the Helmholtz free energy, h is Planck's constant, and the integration is carried out over all DN of the particle coordinates r and momenta p. The pressure can be obtained from the partition function by differentiating it with respect to volume

$$P = -(\partial A/\partial V)_T = kT\,(\partial \ln Z/\partial V)_T.$$

But straightforward differentiation is inconvenient because the volume dependence of the partition function is implicit, appearing only in the integration limits. For simplicity, consider a three-dimensional cubic volume with $V = L^3$. Next, introduce the dimensionless distance variables indicated in **Figure 7**:

$$x = r/L; \quad L = V^{1/3}.$$

$$\vec{x} \equiv \frac{\vec{r}}{L}$$

Old　　　Figure II.7　　　New

Then, the volume dependence appears instead only as a multiplicative factor of V^N in the partition function and in an explicit dependence of the potential energy $\Phi\,(x\,V^{1/3})$ on the volume. Differentiation of the partition function with respect to volume, when expressed in terms of the original variables then yields the usual virial theorem:

$$PV/(NkT) = 1 + \sum\sum [(r_{ij} \cdot F_{ij})/(DNkT)].$$

This expression has been used for over 30 years in obtaining pressure from many-body simulations. The theorem can be applied to elastic solids by making the *shape* of the volume, as well as the size, variable. To treat solids in a consistent way it is convenient to restrict (the center of) each particle to an individual cell, of volume V/N, and to introduce reduced coordinates which span this reduced volume rather than the total volume V. The tensor version of the virial theorem then results:

$$PV = NkTI + \sum\sum r_{ij}F_{ij}.$$

where I is the unit tensor and the double sum includes each pair of particles. The "unit tensor" I has unity for each of its diagonal elements and zero for each off-diagonal element.

But, for nonequilibrium systems, where no analytic expression for a partition function can conveniently be used, this virial theorem is not applicable. In that case, a second, more useful, approach to the virial theorem begins with Newton's atomistic equation of motion $\dot{p} = F$, multiplied by r, summed up over all particles, and averaged over a sufficiently long time τ:

$$\left\langle \sum r\dot{p} \right\rangle_{time} = (1/\tau) \int_0^\tau \sum [d(rp)/dt - (pp/m)]\, dt = \left\langle \sum r_i \cdot F_i \right\rangle_{time};$$

$$-NkTI = -PV + \sum\sum r_{ij}F_{ij}.$$

The pressure tensor P arises by separating the total force on each particle into "external" (wall) forces described by the pressure tensor and "internal" (particle) forces described in terms of microscopic interparticle forces. We work out the external forces for the simplest case, a cubic box. This involves no loss of generality because, if the box is sufficiently large, such sums become independent of box shape. The external forces acting on the x face of a cubic box are $\pm P_{xx}V^{2/3}$, $\pm P_{yx}V^{2/3}$, and $\pm P_{zx}V^{2/3}$. It is necessary to assume that the wall forces are relatively short-ranged, so that the x coordinates of particles interacting with the corresponding pair of faces normal to the x axis differ by $V^{1/3}$. In that case the summed contributions are $-P_{xx}V$, $-P_{yx}V$, and $-P_{zx}V$.

The relationship between the internal and external forces is a tensor equation. The tensors contain sums involving the interparticle forces and the particle velocities. The equivalence, F (on i due to j) $= -F$ (on j due to i) $= F_{ij}$, has been used to combine the two terms $r_i F_{ij}$ and $r_j F_{ji}$ to which the force F_{ij} makes a contribution. The second-line equality, which introduces the temperature T, follows only at thermal equilibrium, where T is isotropic and well-defined. In that case the tensor $\langle \sum pp/m \rangle$ is diagonal, and the three diagonal elements have time-averaged values equal to NkT.

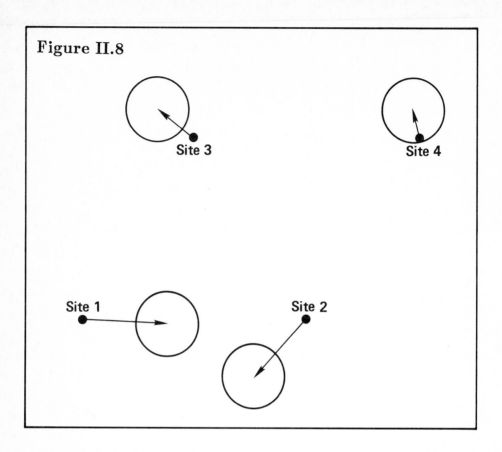

Figure II.8

This time-averaging derivation can be modified in a conceptually important way, for solids. This involves introducing coordinates R fixed at the lattice sites and corresponding displacement coordinates dr such that $r = R + dr$. In **Figure 8** the sites are indicated by filled circles and the displacement vectors dr by arrows. If we multiply the atomistic equation of motion $m\ddot{r} = F$ by the *lattice* coordinates R rather than the instantaneous coordinates r of each particle—and this is unambiguous for solids—then the result for the pressure looks very much the same. But the dynamical variable r_{ij} is replaced by R_{ij} and the velocity terms on the lefthand side do not appear because $\dot{R}_{ij}$ is zero. Thus an alternative, but still exact, form for the solid-phase virial theorem is

$$\sum \sum R_{ij} F_{ij} = PV.$$

This lattice-coordinates version of the virial theorem, with $R_{ij} = R_i - R_j$, can also be derived from the *solid*-phase partition function by using the special particle displacement coordinates dr just defined. In **Figure 9** three deformation modes for a (periodic) volume element containing two particles shown. The lattice coordinates, shown as filled circles, R, are carried

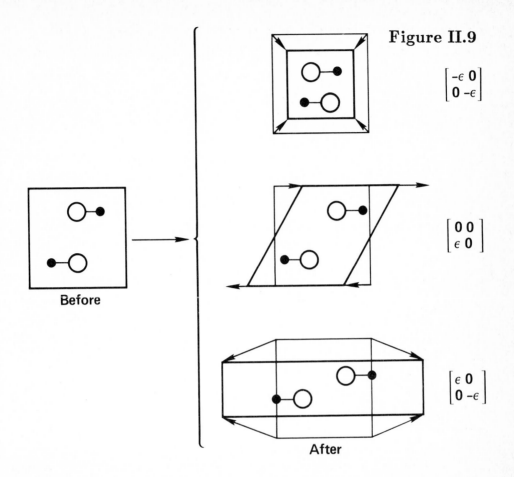

Figure II.9

$$\begin{bmatrix} -\epsilon & 0 \\ 0 & -\epsilon \end{bmatrix}$$

$$\begin{bmatrix} 0 & 0 \\ \epsilon & 0 \end{bmatrix}$$

$$\begin{bmatrix} \epsilon & 0 \\ 0 & -\epsilon \end{bmatrix}$$

Before

After

along with the deformation while the displacement coordinates, shown in **Figure 9** as horizontal lines joining the open-circle "particles" to their lattice sites, dr, are unchanged. These displacement coordinates are tied to lattice positions, $r = R + dr$. Then, in the differentiation of the Helmholtz free energy, only the R coordinates depend upon volume or shape. This solid-phase virial theorem has the advantage that it could be applied to a nonequilibrium steady state in which the external pressure forces and the temperature are constant. But the time-averaging step is particularly inconvenient in treating transient nonequilibrium problems.

A third, and still more useful, virial theorem can be obtained by returning to the fundamental definition of pressure as a momentum flux. In the Lagrangian frame a particle in the volume $V = L_x L_y L_z$, moving a distance $(p_x/m)\,dt$ and carrying with it some property β contributes $(p_x/m)\beta/V$ to the x component of the flux of β. To see this, imagine sampling this flux component in a uniform way, with a small element of area $dy\,dz$ oriented perpendicular to the x direction. What is the probability that this element will intersect the trajectory through which a particle moves during dt? In the x direction the probability of overlap is $(p_x/m)\,dt/L_x$, where

L_x is the box length in the x direction. In the yz-plane at any x the probability of overlap is just $dy\,dz/L_yL_z$. Thus the *flux* of p, the probability of observing the flow of β divided by the time dt and by the area L_yL_z, is $(p_x/m)\beta/V$.

To illustrate the calculation of such a flux we imagine a shear flow in which the x velocity component varies with y, so that the transverse momentum flux, $P_{xy} = P_{yx}$, is nonzero. Consider the contribution of a particle to $P_{yx}V$, the flow of x momentum p_x in the y direction. The quantity being carried, p_x, intersects a plane perpendicular to the y direction with probability $(dz\,dx/L_zL_x)\big[(p_y/m)\,dt/L_y\big]$, for a momentum flux contribution $p_xp_y/(mV)$. The *total* pressure tensor for a volume V includes the corresponding sum over all particles,

$$\sum(pp/m) = P_{Kinetic}V \subset PV.$$

But momentum can also be transferred from Particle i to Particle j through the interparticle forces exerted by i on j and j on i, F_{ji} and $F_{ij} = -F_{ji}$. This momentum transfer can be visualized as a direct connection between the two interacting particles $r_{ij} = r_i - r_j$. (If we wanted to derive the result for a solid we could alternatively use the separation vector linking two lattice sites, R_{ij}.) Then the probability of intersection of the ij vector with an area $dz\,dx$ perpendicular to the y axis is (the absolute value of) $y_{ij}\,dz\,dx/V$.

The complete instantaneous momentum flux has the form

$$PV = \sum(pp/m) + \sum\sum r_{ij}F_{ij}.$$

The expression is clearly symmetric, with $P_{ij} = P_{ji}$. Only six of the nine elements of this tensor are independent (three of the four in two dimensions). In the alternate expression involving the lattice coordinates R_{ij} this symmetry is lost. A very similar complexity results if we consider molecules composed of atoms and imagine that the momentum of each molecule is located at its center of mass. The best feature of this third and most useful, direct, derivation of the virial theorem from momentum flux is that the unpleasant and impractical time-averaging, used to avoid the $\sum r\dot{p}$ term, as well as the fluctuations in the external forces, is not required. The third and last instantaneous form of the virial theorem applies equally well far from equilibrium.

In most molecular dynamics simulations relatively short-ranged forces are used. It is apparent that the minimum-image method for calculating the energy, in which each Particle i interacts with the nearest image of Particle j, leads in a natural way to the realization that the r_{ij} appearing in the virial theorem are minimum-image separations.

In computer simulations this last expression for the pressure tensor is the most useful one. An analogous form for the heat flux vector can be derived by writing down the equation for the rate at which the energy of Particle i changes, $\dot{E}_i$, and multiplying by either r_i or R_i. If an external source of heat produces a heat flux Q by interaction with boundary particles, then the resulting theorem, the "Heat Theorem", is

$$QV = \sum p_i E_i/m_i + \sum\sum r_{ij}\left[F_{ij}\cdot(p_i+p_j)/(2m)\right],$$

where again the double sum is over all $N(N-1)/2$ pairs of particles in the volume V.

If we multiply by R_i rather than r_i, the first sum disappears, because $\dot{R}_i$ vanishes and the double sum contains R_{ij} rather than r_{ij}:

$$QV = \sum\sum R_{ij}\left[F_{ij}\cdot(p_i+p_j)/(2m)\right].$$

Just as with the virial theorem, an instantaneous derivation of the heat theorem can be based on a physically-based microscopic calculation. The kinetic contributions come from the rate at which particles carry energy, with each single-particle contribution multiplied by the probability that the corresponding trajectory intersects a sampling plane. Potential contributions to the heat flux occur whenever two moving particles interact in such a way that one particle transfers a part of their joint energy to the other particle.

The nature of this energy transfer is most obvious in a collision between two hard spheres, with one at rest before collision, and the other at rest after collision. In such a collision the total energy of the pair is transferred through a distance σ, the hard-sphere diameter, at the instant of collision.

II.D Elastic Constants

"Elastic constants" are equilibrium thermodynamic properties which describe the response of the pressure tensor to strain. These strains describe the stretches in the x, y, and z directions, as well as the changes in angles at the corners of a slightly-deformed cube. Thus the deformation of a three-dimensional solid can be described by specifying these six independent strains. In a two-dimensional solid there are only three independent strains, the stretches in the x and y directions and the change in the angle at the corner of a slightly-deformed square. **Figure 9** illustrated these deformations. If the volume has already been specified as an independent variable the number of independent strains is reduced, from 6 to 5 in three dimensions, and from 3 to 2 in two dimensions.

At equilibrium, from either the microscopic or the macroscopic viewpoint, reversible elastic strain deformations can be carried out either isothermally or isentropically, so long as the change is sufficiently slow. Thus two different types of elastic constants—isothermal and adiabatic—can be measured for any particular choice of geometric deformation. In either case the kinetic energy, which measures the microscopic temperature, must remain in equilibrium with the potential energy throughout the deformation. In the adiabatic case temperature generally varies with deformation, but in such way that the energy change is solely the result of reversible "work terms" of the form $-VP : \nabla u$.

Adiabatic sound waves result from macroscopic sinusoidal deformations, with wavelengths sufficiently long that heat conduction between the compressed and dilated parts of the crystal can be ignored. Thus the calculated *microscopic adiabatic* elastic constants correspond to the moduli which describe the propagation of *macroscopic* sound waves.

There seems to be no direct dynamical derivation for the elastic-constant formulae. Microscopic expressions for the elastic constants follow from differentiation of the microscopic internal energy with respect to strain at constant entropy or the Helmholtz free energy with respect to strain at constant temperature. In both cases the first derivative gives the usual pressure tensor. The *second* derivative produces averages of fluctuations of pressure-tensor components. These so-called "fluctuation terms" arise from the changes in relative probability of a system's states brought about by the deformation itself.

The reason for the extra terms can be understood by applying time-dependent perturbation theory. In *first* order, the theory shows that a system does not change state when it is deformed reversibly. Thus the pressure, $P = -(\partial E/\partial V)_S$, can be calculated by averaging $-(dE_{state}/dV)$ over all of the quantum (or classical) states of an elastic solid. In *second* order the state populations change. Thus the isentropic bulk modulus, $-V(\partial P/\partial V)_S$ *cannot* be calculated by so averaging the adiabatic *second* derivative, $(d^2 E_{state}/dV^2)$.

Just as in deriving the virial theorem, there *are* alternative expressions for the elastic constants corresponding to the two kinds of deformation considered in our pressure-tensor derivation, deforming the individual atomic coordinates r and deforming the underlying lattice-site coordinates R.

In computer simulations the elastic constants can be determined by using the equilibrium fluctuation formulae. Alternatively, nonequilibrium simulations of deformation can be used. If the deformation is slow the response can be reversible and either adiabatic or isothermal. If it is fast the viscoelastic respose can be measured. Artificial deformations can be carried out so rapidly that the structure has no chance to relax. Then the so-called "infinite-frequency" elastic constants are measured. Simulations of elastic response over the whole frequency range from the zero-frequency adiabatic soundwave moduli to the infinite-frequency ones can be carried out using the nonequilibrium methods described in Chapter IV.

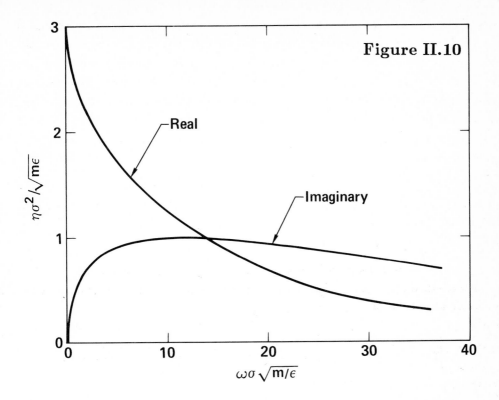

Figure II.10

Typical results for a dense liquid, taken from Denis Evans' work, are sketched in **Figure 10**. A typical single-particle vibrational Einstein frequency would correspond to the middle of the frequency range shown in the **Figure**. At that frequency the response of a fluid to a sinusoidal strain is mainly elastic (labelled "imaginary" in **Figure 10**) rather than viscous (which is labelled "real" in that **Figure**).

Simulations become difficult outside the regime of small-amplitude homogeneous deformation. Exploratory work on high-amplitude deformation and plasticity has shown that the flow response is sensitive to both geometry and size. This work suggests strongly that new and fundamental understanding of deformational constitutive relations will result from increased computational power.

II.E Number-Dependence

The number-dependence of small-system results needs to be understood in order to make the most efficient use of computer time. It is generally best to study the properties of several systems of different sizes rather than expending the entire computational effort on one large-sized system. The variation with number found in this way can be an accurate guide to extrapolation. By number-dependence we mean the way in which a thermodynamic or hydrodynamic system property depends upon the size of the system. Such a property must first be defined in a clear way. There are several different ways in which mechanical properties such as pressure can be measured. Not only can different methods produce different values, but also the corresponding fluctuations can differ. The differences vanish for sufficiently large systems, and the fluctuations vanish, but for finite systems the number-dependence is sometimes the largest source of uncertainty in a computer simulation of macroscopic properties. For this reason it is necessary to choose carefully the method through which a particular many-body property will be determined.

For gases and for solids the number-dependence of thermodynamic properties can be understood theoretically. Consider, for instance, the canonical partition function for the three hard disks shown in **Figure 11**:

$$Z_3 = (2\pi mkT)^3 \iiint\iiint (1 + 3\,\overset{o}{o\!-\!\!o} + 3\,\overset{o}{o\!\angle\!o} + \overset{o}{o\!\triangle\!o})\, dr^3/(3!\, h^6).$$

The diagrams represent the Mayers' expansion of the Boltzmann factor $e^{-\Phi/(kT)}$ into a product of terms:

$$\left[e^{-\Phi/(kT)}\right] = \prod\left[1 + \left(e^{-\phi/(kT)} - 1\right)\right],$$

in which lines joining particles together represent factors

$$\left[e^{-\phi/(kT)} - 1\right].$$

Two separate types of number-dependence can be seen in this simple three-particle example. The 3 appearing in the second term of Z_3, for instance, corresponds, in the general N−particle case, to the number of *pairs* of particles in the system, $N(N-1)/2 = 3$. The presence of an $N-1$, rather than N, produces a number-dependence, in the free energy per particle, of order $1/N$. Likewise, the $N!$ in the denominator introduces a different number-dependence in the entropy and free energies per particle, of order $(\ln N)/N$. This dependence can be seen by using Stirling's expansion to the factorial:

$$(1/N)\ln N! = \ln(N/e) + (\ln N)/2N + (\ln 2\pi)/2N + O(N^{-2}).$$

The same logarithmic dependence arises again in the two-dimensional solid. As a series of progressively larger crystals is examined the low-frequency end of the vibrational spectrum gradually dominates the high frequencies in determining the mean-squared displacement of the particles,

$$\langle \delta r^2 \rangle = DkT \langle 1/\omega^2 \rangle / m.$$

In two dimensions the linear dependence of the frequency distribution, proportional to ω for low frequencies, leads to a logarithmic divergence of the mean-squared displacement, proportional to $(kT/\kappa)\ln N$, where κ is a force constant. The gradual growth of the mean-squared displacement

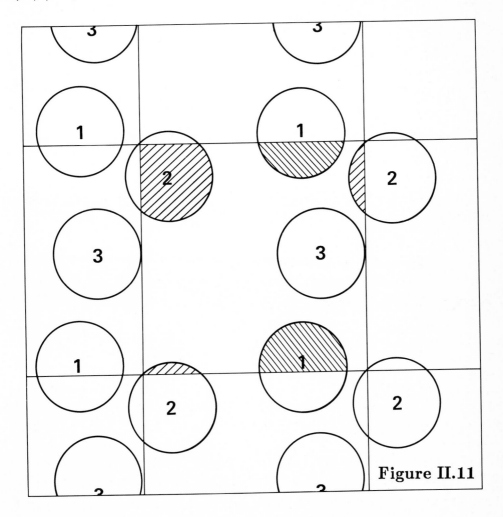

Figure II.11

for a series of N-particle nearest-neighbor triangular-lattice crystals is shown in **Figure 12**. The slope can alternatively be calculated directly from elastic theory, and agrees well with the numerical data.

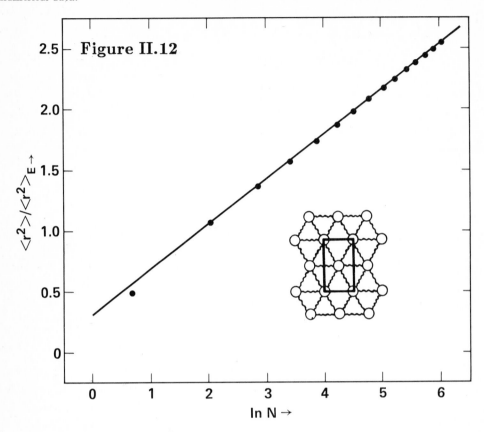

Figure II.12

Both kinds of number-dependence illustrated here, $1/N$ and $(\ln N)/N$, are typically present in the mechanical and thermal properties of fluids and solids. It is usual to express the dependence of hard-particle macroscopic properties on the volume relative to the close-packed volume. For disks and spheres of diameter σ, these volumes are

$$A_o = (3/4)^{1/2} N \sigma^2 \quad \text{and} \quad V_o = (1/2)^{1/2} N \sigma^3,$$

respectively. To illustrate these number-dependent effects in a quantitative way consider the following Table of compressibility factors PV/NkT for three hard disks at a density one-fourth the close-packed value, with $V = 4 \times 3 \times (3/4)^{1/2} \sigma^2$:

	Thermodynamic	Monte Carlo	E constant	K constant
PV/NkT	1.68	1.49	1.74	1.99

The first entry in the Table gives the large-system "Thermodynamic" value. The "Monte Carlo" compressibility factor describes the result of an analytic evaluation of the three-disk partition function equivalent to an exact canonical-ensemble Monte Carlo simulation. The last two values, likewise calculated analytically, correspond to two different kinds of three-disk molecular dynamics calculations, isoenergetic and isokinetic. The distinction between isoenergetic and isokinetic molecular dynamics may appear strange for hard disks, but hard-disk dynamics only makes sense as a limiting case, with the range in which the forces act approaching zero. In this limit, two different results are obtained, one for isoenergetic (Newtonian) dynamics and one for isokinetic (Gaussian) dynamics. Both results approach the thermodynamic limit as the number of particles is increased from three toward infinity, but exhibit deviations of order $1/N$ from this limit in intensive thermodynamic properties.

The situation for transport properties is similar. The Newtonian viscosity for a periodic low-density system of two hard disks,

$$\eta = 0.166 \, (mkT)^{1/2}/\sigma,$$

is about half the infinite-system value. Similarly, the heat conductivity for a moderate-density three-disk system is lower than the large-system value by about a factor of three. This last comparison is at present a little uncertain because the three-disk conductivity appears to vary logarithmically with the strength of the current, as discussed at the end of Chapter IV.

Small systems often provide very interesting and suggestive results in return for relatively little computational effort. A periodic system of two hard disks is a good example. The canonical partition function for two disks or spheres can be worked out analytically at any density up to the maximum, at which the area A equals the close-packed area A_o. If boundaries are chosen as shown in **Figure 13**, so that the two disks can make a "triangular" (close-packed) lattice at high density (so that one side of the rectangular periodic box is $3^{1/2}$ times the length of the other), then an interesting equation of state, shown as a solid line in **Figure 13**, results. It has a loop resembling a van der Waals loop linking a diffusionless solid phase to a fluid phase. This is a simple example illustrating the possibility of finding phase transitions, singularities in free-energy derivatives, in finite, even very small, systems.

It is also shown in **Figure 13** that the pressure and volumes at which this two-disk melting transition occurs lie within 10% of the large-system values indicated by the dashed line. Similar close agreement is found for the three-dimensional hard-sphere analog. The melting transitions found for two-dimensional hard disks and three-dimensional hard spheres were early successes of equilibrium molecular dynamics and Monte Carlo simulations. The phase transition found in two and three dimensions occurs in other numbers of dimensions too. See the paper by Michels and Trappeniers cited in the Bibliography for a discussion of the hard hypersphere melting transition.

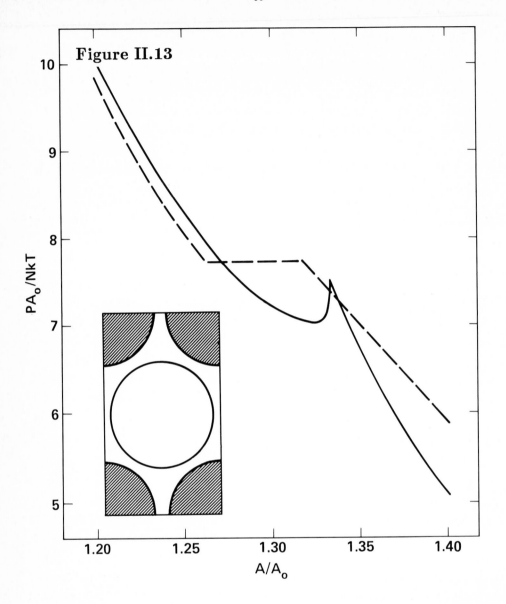

Figure II.13

Solids also exhibit number-dependence in their constitutive properties. Again, these effects are usually of order $1/N$ and $(\ln N)/N$. Both of these dependences can be illustrated and understood by considering the simplest prototype of a crystal, a periodic one-dimensional chain of Hookes'-Law oscillators. We show a three-particle specimen in **Figure 14**. We choose the spring constant, interparticle spacing, and mass all equal to one. With these units a two-particle chain has a single vibrational frequency of $4^{1/2} = 2$. Because the two particles would move in opposite directions the force each would feel is *twice* that which would result if the other were fixed. There is a *second* factor of two because there are *two* springs, not just one, in a periodic system. For the three-particle chain there are two degenerate modes with frequencies $3^{1/2}$, in

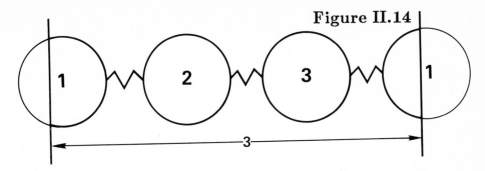

Figure II.14

which one particle moves with an amplitude twice the magnitude, and opposite in sign, of the others. In a four-particle chain we have the same vibration that occurs in the two-particle chain as well as two more degenerate modes, with frequencies $2^{1/2}$, which occur when two non-neighboring particles are fixed and the remaining two move in opposite directions.

From these examples we see that the products of the nonzero frequencies in the 2,3, and 4−particle periodic chains are respectively 2,3, and 4. Thus we correctly guess that the Nth root of the fixed-center-of-mass vibrational partition function for an $N+1$ particle chain with N vibrational modes,

$$\left[Z_{N+1}^{vibrational}\right]^{1/N} = \langle\, kT/h\nu\,\rangle = \langle\, kT/h\nu\,\rangle_\infty \left[(N+1)^{-1/N}\right],$$

lies below the infinite-system limit by a factor of $(N+1)^{1/N}$, so that the small-system free energy per mode exceeds the limiting thermodynamic-system value by $kT\left[\ln(N+1)\right]/N$.

II.F Results

Molecular dynamics simulations began over 30 years ago. Since then a vast body of gas, fluid, and solid results has been generated. These data have been invaluable in developing theories and models linking microscopic mechanisms to experimental data. Some of the early work, exemplified by Rahman's efforts, was motivated by the availability of distribution functions from xray and neutron data as well as Kirkwood's efforts to calculate these functions theoretically. As it became clear that computers could simulate the structure and motions of real fluids and solids, the work became more quantitative, elaborating the thermodynamic properties for simple potentials, over a wide range of conditions, to accuracies of a percent, or better.

By combining thermodynamic properties from different phases, "phase diagrams" were mapped out for a variety of simple pair potentials. The simplest of these is the hard-sphere phase diagram shown in **Figure 15**. The diagram divides all possible equilibrium states for N hard spheres in a volume V at a temperature T into three classes: on the left, fluid; on the right, solid; in between, fluid-solid mixtures. Hard spheres, like hard disks, exhibit only two phases, fluid and solid. The fluid phase transforms at high density into the solid phase. At expansions $(V/V_o) - 1$, relative to close packing, between 35% and 50% an equilibrium hard-sphere system contains both phases. In principle phase diagram calculations of this kind are relatively easy to carry out. If only single-phase calculations are included, first compute the pressure (using the virial theorem) and temperature $\langle p^2 \rangle/(Dmk)$ as functions of energy and volume, $P(E,V)$ and $T(E,V)$ for each phase. Then, use thermodynamic relations to find the Gibbs' free energy, $G = E + PV - TS$ as a function of P and T for each phase. From this the phase diagram follows easily.

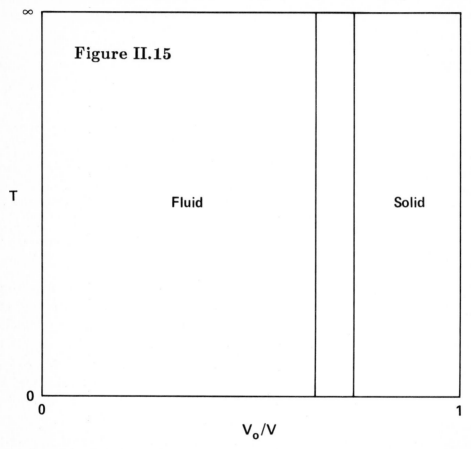

In practice there can be difficulties with the free-energy approach due to the persistence of metastable states. This metastability difficulty can be minimized by using external fields to stabilize phases or to promote the conversion process linking pairs of phases. The essential idea is to develop a reversible computer process for the phase transformation, to which thermodynamics can be applied. Such studies have shown that not only do hard disks and spheres exhibit a first-order melting transition, but also that these transitions can be used as prototypes for melting in real systems. An alternative to the free-energy approach is to simulate a system in which two or more thermodynamic phases coexist, thereby automatically satisfying the requirements of a common pressure, temperature, and chemical potential.

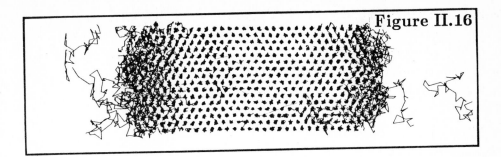

Figure II.16

A simulation carried out at the two-dimensional Lennard-Jones triple point is illustrated in **Figure 16**.

As reliable thermodynamic data became available for different pair potentials the value of corresponding-states approaches, through which properties of different systems are correlated, was established. The success of corresponding-states or perturbation theories of dense fluid properties can be understood in terms of old variational principles due to Gibbs. He pointed out that the distribution function in phase space characterizes each Hamiltonian and that the use of a distribution function for one Hamiltonian to calculate thermodynamics for a different one (the essence of perturbation theory) always leads to a positive error in the Helmholtz free energy. This principle has been developed into an operational model capable of predicting gas and fluid properties for simple materials with high accuracy. A related theory of solid properties has also emerged.

Less is known away from equilibrium, because the problem is a harder one. It has been established that the properties of low-density gases not too far from equilibrium can be described quantitatively by solutions of the Boltzmann equation. A successful combination of the Liouville probability propagation at constant density with the spreading described by Kolmogorov's entropy—actually a *rate* of entropy production—hasn't yet been carried out. But there are indications that transport properties can also be described by corresponding states relations. For instance, we will see in Chapter IV that the thermal conductivity for fluids composed of particles interacting with simple force laws can be found within about ten percent by relating it to the excess entropy. This relationship between conductivity and entropy can be understood by relating both properties to the Einstein vibrational frequency. In view of the difficulties standing in the way of an exact theory, the computer simulations are essential to developing understanding of nonequilibrium corresponding-states relations.

Bibliography for Chapter II

Free volumes for hard disks are described in a paper (rejected by Molecular Physics) by W. G. Hoover, N. E. Hoover, and K. Hanson, "Exact Hard-Disk Free Volumes", Journal of Chemical Physics **70**, 1837 (1980).

Monte Carlo calculations of elastic constants for Lennard-Jones crystals appear in A. C. Holt, W. G. Hoover, S. G. Gray, and D. R. Shortle, "Comparison of the Lattice-Dynamics and Cell-Model Approximations with Monte-Carlo Thermodynamic Properties", Physica **49**, 61 (1970).

The viscoelastic fluid-phase results in Section D are reproduced from D. J. Evans, "Rheological Properties of Simple Fluids by Computer Simulation", Physical Review **A23**, 1988 (1981).

The number-dependence of the equation of state for systems of hard disks and hard spheres is discussed in a paper by W. G. Hoover and B. J. Alder, "Studies in Molecular Dynamics. IV. The Pressure, Collision Rate, and Their Number-Dependence for Hard Disks", Journal of Chemical Physics **46**, 686 (1967).

The number-dependence of the entropy is discussed in "Number-Dependence of Small-Crystal Thermodynamic Properties", by W. G. Hoover, A. C. Hindmarsh, and B. L. Holian in the Journal of Chemical Physics **57**, 1980 (1972). A formalism, the cell-cluster theory, well-suited to crystal defect free-energy calculations is discussed in a series of papers by Dale Huckaby, Cesar M. Garza, Howard S. Carman, and Robert H. Kincaid, all in the Journal of Chemical Physics: "Derivation of the Cell-Cluster Theory for Harmonic Solids", **65**, 607 (1976); "Effect of a Pair of Vacancy Defects on the Free Energy of Harmonic Cystals", **73**, 1923 (1980); "Cell-Cluster Calculations of the Entropy of Cracked Crystals", **75**, 4651 (1981); "Harmonic Surface Entropy of Noble Gas Crystals", **78**, 2598 (1983).

The phase diagrams for three-dimensional inverse-power repulsive potentials are estimated in a paper by W. G. Hoover, M. Ross, K. W. Johnson, D. Henderson, J. A. Barker, and B. C. Brown, "Soft-Sphere Equation of State", Journal of Chemical Physics **52**, 4931 (1970).

J. P. J. Michels and N. J. Trappeniers "Dynamical Computer Simulation on Hard Hyperspheres in Four- and Five-Dimensional Space", Physics Letters A **104**, 425 (1984) describe the calculation of phase diagrams for four- and five-dimensional hyperspheres.

An accurate equation of state can be calculated for particles interacting with pairwise-additive forces using "A Perturbation Theory of Classical Equilibrium Fluids", by H. S. Kang, C. S. Lee, and T. Ree, and F. H. Ree, Journal of Chemical Physics **82**, 414 (1985). For solids a similar approach can be followed: H. S. Kang, T. Ree, and F. H. Ree, "A Perturbation Theory of Classical Solids", in the Journal of Chemical Physics **84**, 4547 (1986).

A corresponding states treatment of fluid-phase heat conductivity based on computer simulations is discussed by R. Grover, W. G. Hoover, and B. Moran, "Corresponding States for Thermal Conductivities *via* Nonequilibrium Molecular Dynamics", Journal of Chemical Physics **83**, 1255 (1985). See Section I of Chapter IV of these notes.

Farid Abraham kindly provided a preliminary draft of his comprehensive and stimulating review, "Computational Statistical Mechanics, Methodology, Applications, and Supercomputing", to appear in Advances in Physics (1986). The last **Figure** in this Chapter, the last **Figure** in Chapter I, and the first **Figure** in Chapter IV were all reproduced from originals in this review article.

III. NEWTONIAN MOLECULAR DYNAMICS FOR NONEQUILIBRIUM SYSTEMS

III.A Limitations of the Newtonian Approach

Conservative Newtonian mechanics can be used to treat a variety of nonequilibrium problems involving the approach to equilibrium of isolated systems or the mechanical coupling of systems to external energy sources. The expansion of a system into a vacuum, the equilibration of a nonequilibrium velocity distribution, and the relaxation of initially-excited modes are all examples of isolated systems exhibiting interesting transient behavior. Systems treated with Newtonian mechanics need not be isolated. A system can be compressed by using a rigid wall, converting the mechanical work done by the wall into internal energy. But without formulating a thermostat, or some method for equilibrating thermally with an external reservoir, only *mechanical* problems can be treated with Newtonian mechanics. In particular, problems involving steady nonequilibrium states *always* require the extraction of heat by an external force that cannot be formulated using Newtonian mechanics. For this reason many of the applications to nonequilibrium problems require new equations of motion. We will come to the implementation of thermostats and modifications of the Newtonian approach in Chapter IV. In this chapter we concentrate instead on what can be done with gases, liquids, and solids obeying Newton's equations of motion.

III.B Gases: Boltzmann's Equation

Whenever the density is *low* enough that successive binary collisions are uncorrelated and the collisions can be considered to occur between particles at the same spatial location, but is still *high* enough that the dynamics is not dominated by the boundary conditions, Boltzmann's gas-phase equation

$$df/dt = \partial f/\partial t + v\, \partial f/\partial r + F\, \partial f/\partial p = (df/dt)_{collisions}$$

can be used to describe the corresponding nonequilibrium systems. This equation is the exact Liouville equation until the collision term is approximated to make solutions possible. Boltzmann used a quadratic approximation to the collision term. We will use instead the even cruder *linear* relaxation-time approximation. Here f is the one-body probability density. For three-dimensional monatomic particles this probability density has to be integrated over three space and three velocity coordinates (momenta could equally well be used) to correspond to a (dimensionless) probability. Thus the units of f are either $[\text{seconds}^3/\text{meters}^6]$ or $[(\text{seconds/kilograms})^3/\text{meters}^6]$.

Solutions of the Boltzmann equation incorporating gradients of mass, momentum, or energy lead to numerical estimates of the diffusion, viscosity, and conductivity coefficients in terms of the interparticle forces. John Barker and his coworkers have carried out a careful comparison of the transport properties for the rare gases, from the Boltzmann equation, using a force-law fitted only to equilibrium properties. The agreement was very good.

In deriving the Boltzmann equation for the *one*-body distribution function from the *many*-body Liouville equation, it is necessary to assume that the rate at which collisions occur between particles with two different momenta, p_1 and p_2, is proportional to the numbers of such particles at $r, f(r, p_1, t)$ and $f(r, p_2, t)$. This statistical assumption makes the righthand side of the Boltzmann equation irreversible, unlike the lefthand side (which describes reversible collisionless streaming in phase space) and unlike the Liouville equation. This qualitative difference between the underlying reversible equations of motion and the irreversible equation for the probability density evolution seems paradoxical and has triggered many attempts to "understand" the source of the irreversibility.

As we saw in Chapter I, Alder and Wainwright studied the approach to equilibrium of 100 hard spheres, all moving with the same initial speed, but in different directions. After one collision such spheres can have any kinetic energy between zero and twice the initial energy. After two collisions one sphere could end up with an energy three times the mean, and so on. **Figure 1** outlines a sequence of two such transverse collisions which are maximally effective in transferring energy between colliding particles.

Figure III.1

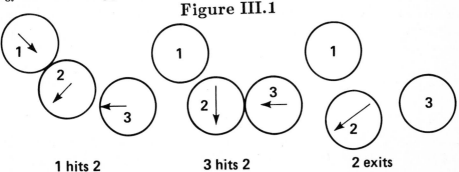

| **1 hits 2** | **3 hits 2** | **2 exits** |

Alder and Wainwright verified that Boltzmann's description of the approach to equilibrium (through following the time dependence of $\langle \ln f \rangle$) was in excellent agreement with their reversible dynamical simulation. Unless the initial conditions are specially chosen to violate it, Boltzmann's irreversible equation is an excellent description of gas-phase dynamics.

We shall see that Boltzmann's equation even describes very small two- and three-body systems. By using periodic boundaries it is possible to simulate diffusive, viscous, and heat flows with such small, few-particle systems. The Boltzmann Equation can be applied to these problems and, in the two-hard-sphere case, it has a particularly simple form. Both the coordinate space and the momentum space are shown for such a two-particle system in **Figure 2**. The two spheres must move in opposite directions, because the only reasonable value of the total momentum, a constant of the motion, is zero. If the Boltzmann equation description can be applied at all to such a simple system the solution can only correspond to the properties of an *ensemble* of these two-body systems in which the initial coordinates and the directions of motion are chosen from an appropriate statistical distribution.

Figure III.2

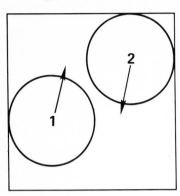

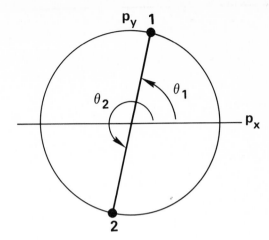

The usual Boltzmann equation, describing the collisions of pairs of particles, is quadratic in f, and therefore nonlinear and difficult to solve. For an ensemble of very simple two-sphere systems, in which all members of the ensemble are restricted to have the same kinetic energy, the Boltzmann equation reduces to a *linear* equation rather than a quadratic one, and becomes relatively easy to solve. The equation is linear because f is the same for both the two colliding particles in a fixed-energy system with fixed center of mass:

$$f(r, p, t) = f(r, -p, t).$$

Given the momentum of either colliding sphere we know the momentum of the remaining sphere. Further, the collision rate is independent of the momentum p, because the *magnitude* of p cannot vary in a constant-energy system. Thus an ensemble of two-body systems can be described by a linear Boltzmann equation. It has the form of the "relaxation-time" approximation often used in treating many-body flows:

$$df/dt = (\partial f/\partial t) + \partial(f\dot{r})/\partial r + \partial(f\dot{p})/\partial p = (f_o - f)/\tau.$$

Here, f_o is the equilibrium post-collision distribution function. We include the possibility that the sum $(\partial\dot{r}/\partial r) + (\partial\dot{p}/\partial p)$ may not vanish. In the general case the time between collisions, τ, depends inversely on the particle speed. It is therefore only in the special case that all particles have the same speed, as in the two-sphere problem, that this approximation becomes exact.

To illustrate the usefulness of the relaxation-time Boltzmann equation let us consider again the problem of Alder and Wainwright, in which a velocity distribution localized to a constant-speed shell approaches the Maxwell-Boltzmann distribution. In the relaxation-time approximation f has the form:

$$f - f_o = [f_{initial} - f_o] e^{-t/\tau}.$$

This approximation provides a useful estimate of the time required to reach equilibrium (a few collision times), but as we saw, the velocities in a many-sphere system are correlated, with the higher speeds filling in more slowly than predicted according to the approximate two-sphere model.

Bird suggested a much better, but computationally more-involved approximation, in which particles are selected for collisions statistically, based on a computed collision rate. In this way the actual particles followed, up to about a million, represent many more particles. This method is specially useful in solving aerodynamic "Knudsen-gas" problems. In such gases the mean free path *is* comparable to boundary dimensions. Thus the particle flow is influenced by the details of the boundary and Navier-Stokes continuum mechanics can be qualitatively wrong.

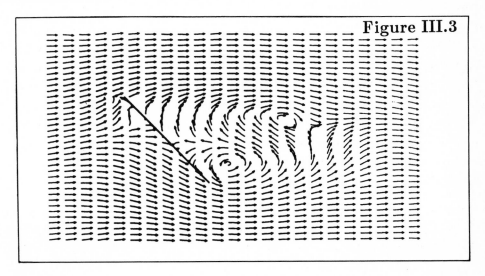

Figure III.3

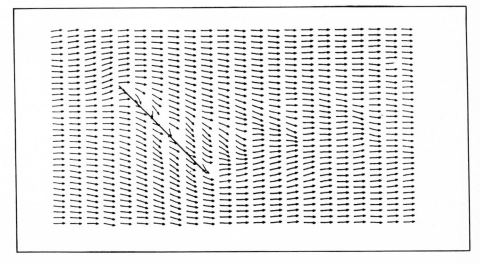

Bird's method has been successfully applied to a host of flow problems involving rarefied gases. Very recently Meiburg carried out a molecular dynamics simulation of the flow of 40,000 hard spheres past an inclined plate. He compared the predictions of Bird's approximation with his own molecular dynamics results. The results of both calculations are shown in **Figure 3**. The top flow field is Meiburg's molecular dynamics result. Each arrow represents the instantaneous average velocity of several dozen hard spheres, all lying within a fixed Eulerian spatial zone. The collisions in Meiburg's calculation are calculated without approximations. The lower flow field is based on Bird's idea of selecting nearby pairs of particles for collision on a statistical basis. The two flow fields are qualitatively different.

Meiburg's calculation shows that the plate inserted into the gas stream generates and sheds vortices when the flow is averaged over zones containing a few dozen spheres each. **Figure 3** shows that these vortices are absent in the calculation carried out using Bird's statistical collisions. The vortices are missing because statistical collisions do not conserve angular momentum. In a "head-on" statistical collision between particles located in different spatial zones, their contribution to the angular momentum changes sign, as indicated in the following **Figure 4**. The effect of these collisions between separated particles is to *lengthen* the effective mean free path, thereby enhancing viscous dissipation and reducing the Reynolds number, a measure of turbulence discussed in Section H of Chapter IV. Meiburg's work is stimulating the development of a hybrid method combining stochastic collisions with angular momentum conservation. It is one of many examples in which the results of the molecular dynamics simulation were surprising and suggestive.

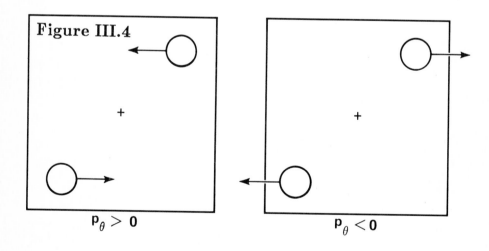

Figure III.4 $P_\theta > 0$ $P_\theta < 0$

III.C Liquids: Shockwave Simulation and Fragmentation

A one-dimensional steady fluid-phase shockwave is one of the simplest and one of the most interesting nonequilibrium systems. Shockwaves are a logical phenomenon to study using molecular dynamics because they are atomic in scale. Within the shockwave the relationship between the longitudinal pressure, parallel to the direction of wave propagation, and volume is a linear one, as we show below. At the same time there are gradients, not just in one property, but simultaneously in *all* of them, pressure, temperature, density, energy, and entropy, for instance. Thus a wealth of far-from-equilibrium effects can be studied by generating shockwaves.

In the shockwave cold material is converted into hot in a steady way, the moving shockwave simply separating two equilibrium fluid states, one hot and moving, the other cold and static, from one another. The usual textbook way of making a shockwave is to push on a fluid with a piston. If this is done with a series of small, successively harder, pushes then the resulting pressure wave has steps in it according to *linear* wave propagation theory. But nonlinearity typically causes the wave velocity to increase with compression. Then the trailing steps catch up and the wave steepens to become a shockwave.

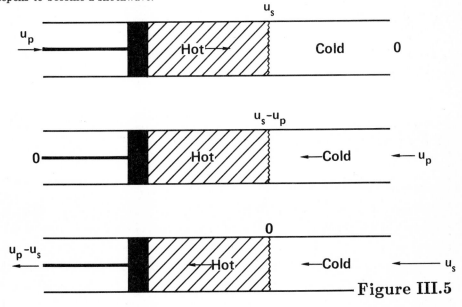

Figure III.5

Figure 5 shows the geometry used to produce a one-dimensional shockwave, viewed in three different coordinate systems. In each view the shockwave is indicated as the transition zone between a cold fluid and a hot fluid. Such idealized steady one-dimensional transformations can be closely approximated experimentally by applying a high pressure to a planar fluid or solid boundary.

In the top view, a piston, moving from the left at speed u_p compresses and heats the static cold material lying to the right. Consider next a coordinate system fixed on the piston. In this middle view the cold material, initially moving to the left at speed u_p stagnates against the fixed piston. In the final bottom view, which provides the simplest theoretical treatment of the problem, the shockwave disturbance is fixed in space with cold material entering from the right and hot material leaving to the left.

In a steady shockwave, generated as is illustrated in **Figure 5**, the description of thermodynamic and hydrodynamic properties is time-independent, and therefore simplest, in the bottom frame, moving with the wave. The variations of the local quantities, density, pressure, energy, and temperature, as shown in **Figure 6**, constitute the shockwave "profile".

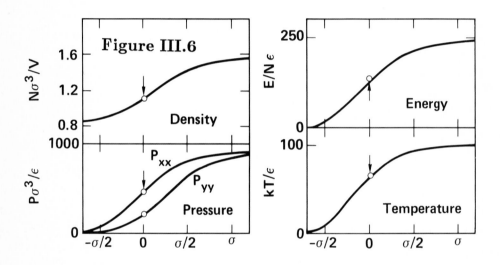

The **Figure** shows very good agreement between the molecular dynamics results—full curves—and a numerical solution of the Navier-Stokes hydrodynamic equations—open circles at the shockwave center. In the bottom fixed-shockwave frame conservation of mass requires that the product of the density and velocity be constant. If the mass flux were not constant, throughout the wave, the density profile would change. The constancy of the product ρu, the steady-flow mass flux, can be used, at the cold and hot ends of the shockwave, to relate the shock velocity u_s and the particle velocity u_p, to the two densities, ρ_0 for the cold material and ρ_1 for the hot:

$$\rho u = \rho_0(-u_s) = \rho_1(u_p - u_s).$$

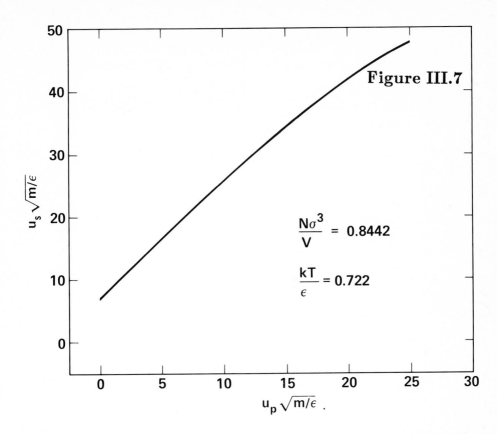

$$\frac{N\sigma^3}{V} = 0.8442$$

$$\frac{kT}{\epsilon} = 0.722$$

Figure III.7

In the absence of phase transitions, and over the range of piston and shock velocities ordinarily observed in laboratory experiments, there is typically a nearly-linear relation between the two speeds, as shown in **Figure 7** for the Lennard-Jones liquid phase initially at the triple point. The relationship linking the shock velocity u_s and the particle velocity u_p covers a range of shockwaves ranging from soundwaves, for which u_p vanishes, up to twofold compression. For a real liquid, such as argon, under the extreme condition of twofold compression, the temperature reached would be of the order of an electron volt.

The flow of momentum must likewise be steady in the frame moving with the wave. There are two contributions to the momentum flux, the comoving flow, in the frame of the fluid, given by the pressure tensor element, P_{xx}, and the additional convective component ρu^2, where u is measured relative to the comoving frame. Thus u is $-u_s$ for the cold material and $u_p - u_s$ for the hot. Conservation of momentum requires that the total momentum flux, given by the longitudinal pressure P_{xx} plus the convective momentum flux $(\rho u)^2/\rho$ be a constant. Applying momentum conservation to the cold and hot end points of the shockwave gives a relation linking the equilibrium pressures P_0 and P_1 of the cold and hot materials:

$$P_{xx} + \left[(\rho u)^2/\rho\right] = P_0 + \left[(\rho u)^2/\rho_0\right] = P_1 + \left[(\rho u)^2/\rho_1\right].$$

Thus, in a steady "one-dimensional" (planar) shockwave, the longitudinal component of the pressure tensor varies *linearly* with volume from the initial value P_0, to the final value P_1, as shown in **Figure 8** for the same Lennard-Jones liquid shockwave. The linear pressure-volume relation is called the "Rayleigh Line".

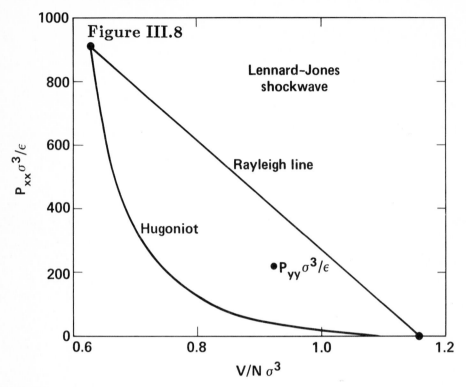

For weak shock waves, the difference between this linear relation and the isentrope is small, with the entropy change ΔS of *third* order in the compression, so that shock waves, for small compressions up to 20%, are very nearly isoentropic. Stronger shockwaves, such as that detailed in **Figures 6 and 8**, attain states far from equilibrium at pressures and temperatures which cannot be achieved by other methods. The locus of pressure-volume states which *can* be achieved from a given initial state (the Lennard-Jones liquid triple point in the case shown in the **Figures**) is called the Hugoniot curve. For any final state P_1, V_1, with energy E_1, the energy change from the initial state P_0, V_0 can be found by integration. The integration is carried out along the appropriate Rayleigh line, the straight line linking the initial and final states, with the result

$$\Delta E = E_1 - E_0 = (1/2)\,(P_0 + P_1)\,(V_0 - V_1).$$

This exact relation, the Rankine-Hugoniot equation, follows from the conservation of energy in the shockwave process.

By measuring the shock and particle velocities of these waves and using the conservation relations the thermodynamic properties far from usual conditions can be measured accurately. Published data for aluminum have been reported at over one gigabar, hundreds of times the pressure at the center of the earth. These pressures are achieved using hydrogen bombs.

We can estimate the physical extent of a one-dimensional shockwave disturbance by setting the generated pressure ΔP_{xx} equal to the product of the viscosity coefficient η and an estimate of the strain rate c/w, where c is the sound speed and w is the shock width:

$$\Delta P_{xx} = P_1 - P_0 = \eta c/w.$$

Using kinetic theory to estimate these quantities gives the results:

$$\eta \sim (mkT)^{1/2}/(\pi\sigma^2); \quad c \sim (kT/m)^{1/2}; \quad P = NkT/V.$$

Thus we find that the shockwidth is of order $V/(N\pi\sigma^2)$, the kinetic-theory mean free path. In a dilute gas, the mean free path is so long, thousands of particle diameters, that the shockwave can scatter light. But for fluids, the mean free path is of the order of one particle diameter. This suggests that the shock transition is localized in condensed matter and can accordingly be studied by molecular dynamics.

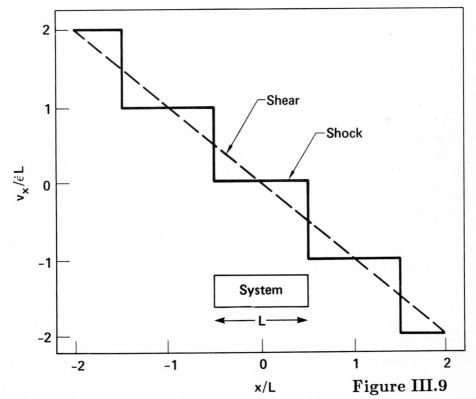

Figure III.9

This has been done by using time-dependent periodic boundaries of the type shown as solid lines labelled "Shock" in **Figure 9**. These boundaries correspond to the compression of an initially quiescent system using two opposed pistons which are periodic, but moving, images of the system itself. The boundaries move at a constant speed ($\pm$ the particle velocity) so that the system is compressed at a constant rate. But adjacent images have different speeds so that the *relative* velocity of a particle, relative to the system image within which it lies, jumps discontinuously, by $\pm v_x$, whenever such a particle crosses an image boundary. In the laboratory frame all particle velocities are continuous. The mean particle velocity within any periodic image of the system has a discontinuity, viewed macroscopically, between shocked material, which has felt the influence of the moving boundaries, and quiescent material, which has not.

If the mean particle velocity is instead chosen initially to vary *linearly* with coordinate, without jump discontinuities, as in the dashed curve marked "Shear" in **Figure 9**, longitudinal viscous flow could be simulated. Because the distortion indicated in the **Figure** is longitudinal, both bulk and shear viscous stresses would result, in addition to the equilibrium increase in pressure associated with a quasi-static reversible compression. We will return to viscous flows in Chapter IV. But in shockwave simulation the velocity instead varies in a series of equally-spaced jumps. The thermodynamic and hydrodynamic profiles describing the shockwave agree semiquantitatively with the predictions of Navier-Stokes continuum mechanics. This continuum model is based on the assumptions that Newtonian viscosity and Fourier heat conductivity are sufficient to describe the nonequilibrium momentum and energy flows. At the highest pressures studied it was found that the shockwidth was a little wider than the continuum prediction. The 30% discrepancy cannot be explained by the dependence of viscosity or conductivity on frequency, wavelength, or strainrate, and is not yet understood.

The molecular dynamics simulations show also that the velocity distribution in the vicinity of the shockwave is very different from the equilibrium one. The Maxwell-Boltzmann relation

$$\langle p_x^4 \rangle = 3 \langle p_x^2 \rangle^2,$$

is a poor approximation near the shock front. The discrepancy with this property of a Gaussian distribution is expressed in terms of the "kurtosis", shown for both the longitudinal and transverse directions in **Figure 10**. In a strong shockwave the longitudinal and transverse temperatures, $\langle p_x^2 \rangle/(mk)$ and $\langle p_y^2 \rangle/(mk)$, can differ by a factor of two. These substantial anisotropies should lead to correspondingly substantial deviations from simple Arrhenius kinetics in shockwave-induced chemical reactions.

Because the shockwave simulations are time-consuming relative to single-phase homogeneous studies, only a few shockwave simulations have been carried out. It is not presently known how well dense-fluid shockwaves or solid-phase shockwaves satisfy corresponding-states relationships. For fluids this could be explored by solving the Navier-Stokes equations for a variety of simple force laws.

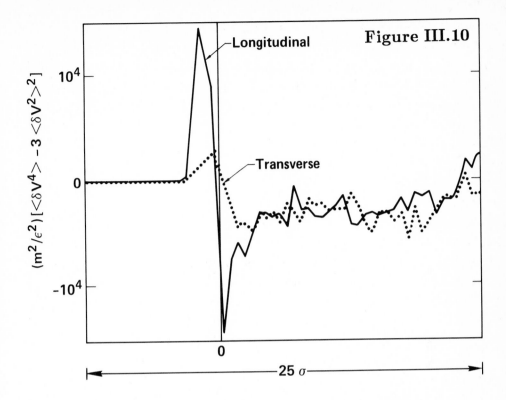

Figure III.10

Another nonequilibrium problem which is similarly free of boundary difficulties is the expansion of a high-pressure fluid or solid into a vacuum. The fluid-phase problem has an application in one design of a fusion reactor in which liquid lithium is used to transfer the heat from the fusion reaction away from the fusion chamber. In this design hot jets of lithium, a few centimeters in radius, absorb the energy of fusion neutrons, and are thereby heated very rapidly, essentially at constant volume, to pressures of a kilobar or so. This high pressure then causes the jets to expand rapidly and fragment. It is important to know whether the energy is primarily kinetic, with large lumps of relatively cold lithium colliding with the walls of the reactor, or thermal and surface, with a quiescent froth of lithium fog harmlessly filling the reactor chamber. Either outcome is consistent with conservation of momentum and energy.

A semiquantitative model, developed by Dennis Grady and elaborated by Lewis Glenn, describes this situation by assuming that the lithium expands in a uniform way, with the outer edge moving at a "jumpoff" velocity $u_j = P/\rho c$, where c is the sound velocity. The local (co-moving) kinetic energy density can then be calculated from the strain rate, u_j/R, to give the available kinetic energy in a fragment of size r. This kinetic energy is of order $r^{D+2}R^{-2}$.

Setting this energy source equal to the surface energy created in the fragmentation process, of order r^{D-1} for a fragment of size r, predicts that the number of fragments varies as the cube root of the total mass available (in either two or three dimensions). This prediction has been confirmed by computer simulations using over 14,000 particles. **Figure 11** shows the conformation of a two-dimensional simulation of such an expansion. The time involved in the simulation is on the order of a few sound traversal times, or about 100,000 time steps.

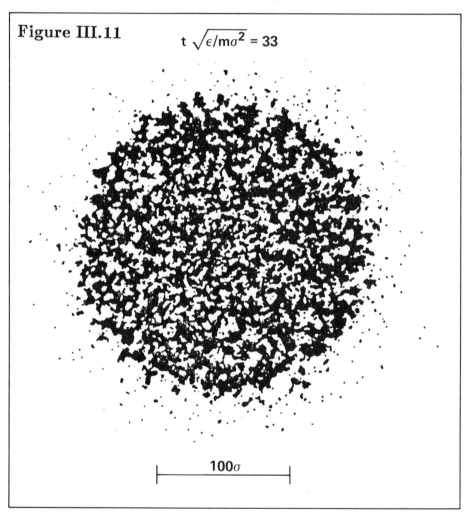

Figure III.11 $t \sqrt{\epsilon/m\sigma^2} = 33$

100σ

III.D Solids: Breakdown of Continuum Mechanics

Many problems in materials science involve defects which are small, either microscopic or mesoscopic in scale. Microscopic diffusion of atoms, both in surface layers and in the bulk, and the mesoscopic motion of the dislocations which are the mechanism for plastic deformation are examples. These problems, which cannot be dealt with using macroscopic continuum mechanics, present a variety of opportunities for stimulating and useful research in molecular dynamics.

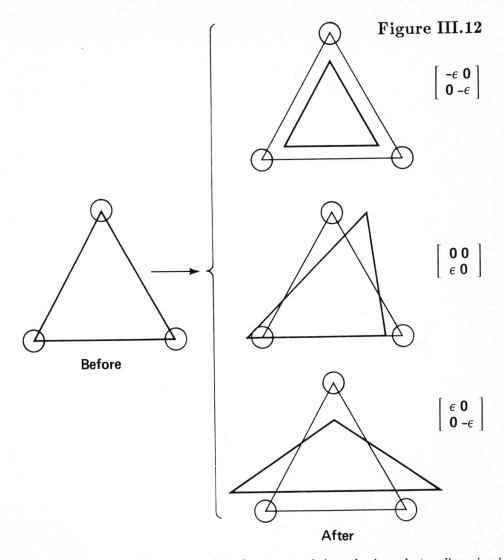

Figure III.12

Before

After

There is an interesting correspondence between atomistic mechanics and a two-dimensional elastic continuum (equivalent to a three-dimensional plane-strain or plane-stress elastic solid in three-dimensional space). A two-dimensional triangular-lattice crystal of Hooke's-Law particles, in which each particle interacts only with its six nearest neighbors, obeying the equations of motion $\dot{p} = F$ can be brought into exact correspondence with an isotropic elastic continuum when the interparticle spacing is small. It is easy to show that a "quasiharmonic" triangular-lattice crystal (one in which energy contributions depend only quadratically on the particle displacement coordinates) is elastically isotropic . This can be done by considering the two shear deformations of a triangle of atoms shown as the lower two deformations in **Figure 12**. Thus the dynamics of the crystal can be described by a bulk modulus and a shear modulus. These same elastic quantities describe the behavior of an elastic isotropic continuum.

The analogy can be made more detailed. If a continuum is divided into equilateral triangular finite elements within which the displacement varies *linearly*, the six coefficients describing this *linear* variation can be determined uniquely from the x and y displacements of the three vertices of each triangle. The dynamics of the vertices can be calculated in two different ways, from the atomistic Hooke's-Law equations or from a finite-element approximation to continuum mechanics. If the time-dependence of the vertex displacements is calculated from the elastic equations in the linear approximation just described, the motion is *identical* to the motion of a Hooke's-Law crystal described using molecular dynamics!

This correspondence shows that the simulation of elastic continuum problems using finite-element methods can produce solutions closely resembling motions seen in atomistic simulations. The primary difference between the macroscopic continuum problems and the microscopic atomistic simulations concerns the dependence of wave velocity on wavelength. In a continuum it is expected that elastic acoustic waves all travel with the appropriate sound velocity, independent of wavelength. This simple behavior is consistent with a *linear* density of vibrational frequencies (quadratic in three dimensions). The exact lattice frequency distribution is shown in **Figure 13**.

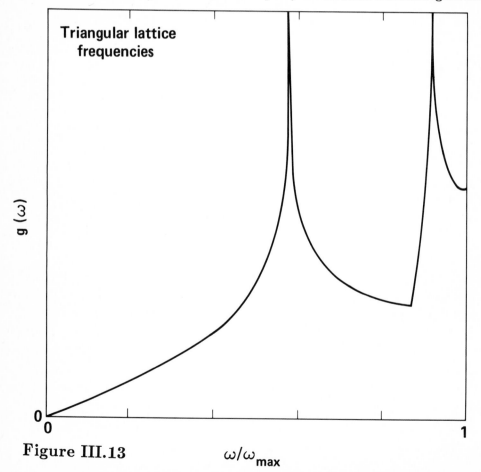

Figure III.13

The **Figure** shows that in the nearest-neighbor triangular lattice, this linear behavior breaks down at a frequency about one third of the maximum, corresponding to a wavelength of six atomic diameters. Accordingly, continuum mechanics can be expected to fail, even in elastic problems, at length scales of order a few atomic diameters.

A similar atomistic-continuum correspondence to the one found in two dimensions, with the triangular lattice, doesn't exist in three dimensions (where the shear modulus is strongly direction dependent in close-packed crystals) or for incompressible fluids (for which it is hard to develop a corresponding atomistic model which avoids expansion or compression of the volume elements). Nevertheless, this correspondence between two-dimensional atomistic and continuum properties is useful in a relatively wide class of problems that involves defects of various kinds: vacancies, dislocations, or cracks, in solids. This is because molecular dynamics is required only to investigate displacements within a few atomic diameters of these defects. Beyond a few diameters continuum elasticity can be used.

Defects are of importance in designing structures because they concentrate stress, and stress causes materials to fail, either by flow or by fracture. The idea of stress concentration is familiar to engineers. If we consider a plate under a uniform vertical longitudinal tensile stress σ, and then bore a small hole in it, of radius a, the maximum tensile stress near the hole will increase by a factor of three, to 3σ. The stress concentration in the vicinity of such a hole is shown in **Figure 14**.

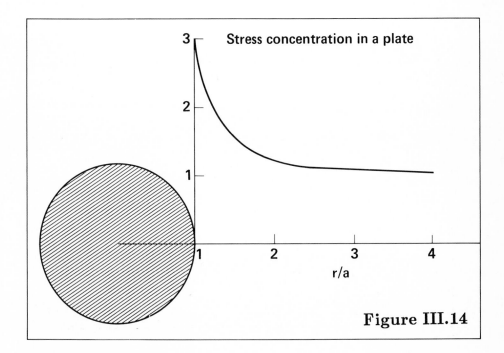

Figure III.14

From the standpoint of safety a circular hole is the *best* case, because it gives the *least* stress concentration. The concentration is *greater* for other shapes, such as ellipses or polygons. The solution for an elliptical hole has an interesting property as the ellipse is deformed to approximate a thin long crack. In that limit the stress concentration *diverges*. This can be seen in another way. If the displacements near a crack tip vary as the square root of the distance from the tip, as would be expected for a parabolic crack tip, then the stresses and strains (proportional to the symmetrized derivatives of the displacements) vary as the inverse square root, in agreement with the analysis based on an elliptical crack.

Thus the structure and motion of cracks cannot be treated by straightforward continuum mechanics. On the other hand, molecular dynamics simulations of crack motion *can* be carried out. The fracture velocities and stress fields correspond relatively well to results obtained experimentally for brittle materials. The energy and entropy associated with static cracks can be calculated easily, but dynamic simulations can lead to highly irregular size-dependent structures. Even for computer experiments, the interpretation of fracture is complicated by the irregular nature of the fracture surface.

In both the microscopic and the macroscopic cases cracks can travel at speeds approaching the transverse sound velocity and can propagate into regions in which the stress is not enough to cause a static crack to move. An example appears in **Figure 15**, taken from Bill Moran's 1983 thesis. He studied the fracture of triangular-lattice crystals by breaking several adjacent nearest-neighbor bonds, forming a "crack", under tensile conditions imposed at the boundaries. Under sufficiently great tensile stresses the stress concentration at the crack tip was sufficient to cause the crack to propagate into unbroken material. In the case shown tapered boundary conditions were used to build a linear decrease of tensile stress with distance into the material. Static simulations indicated that a crack at the location indicated by the arrows could propagate no farther into the crystal. On the other hand a moving crack exhibited "inertia", penetrating into a region with stress some 15% less than would be predicted by a static analysis. The dynamic vibrations of atoms neighboring the crack tip are responsible for this effect.

It is interesting to run a crack-propagation calculation backwards, reducing the stress and watching the crack heal. This healing occurs only with difficulty, moving in a much less steady way than in the more natural (order-destroying) fracture process.

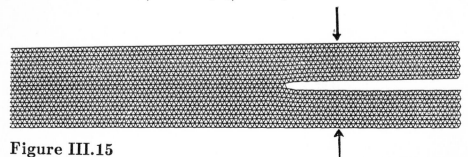

Figure III.15

The dislocations responsible for the plastic flow of rocks and metals are, like cracks, atomistic defects which can also be treated, in an approximate way, using continuum mechanics. Dislocations can be formed by inserting an extra half-plane of atoms into a three-dimensional crystal. In the two-dimensional case the additional particles form an extra row. In the case shown in **Figure 16** *three* dislocations are included in the periodic unit cell. In order to see the dislocations clearly this **Figure** should be viewed obliquely. Compare this three-dislocation **Figure** to the two-dislocation Figure 21 in Chapter IV.

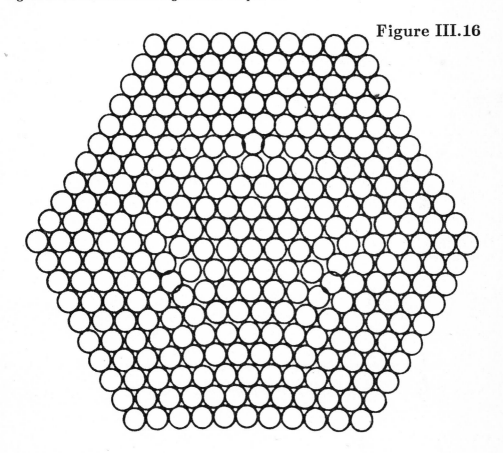

Figure III.16

Because the atoms near a circle of radius r centered on an isolated dislocation will encounter displacements of order one atomic diameter b, the corresponding strain field is of order b/r, more localized than the field associated with cracks. The divergent strain causes the energy density integral, which has an integrand proportional to stress times strain, proportional to r^{-2}, to diverge logarithmically both at small and at large r. Pairs of dislocations interact with a logarithmic tensorial potential and can be treated by a "mesoscopic" dynamics intermediate between molecular dynamics and continuum mechanics. In such simulations the "equation of motion" is the relationship between the dislocation velocity and the local stress tensor.

The parameters required to apply such calculations to a particular material are many, primarily because in real three-dimensional materials the dislocations form loops rather than straight lines, and these loops move locally in a series of jumps and jogs, rather than in synchrony.

These fracture and flow problems are difficult to treat from a realistic point of view because forces, impurities, and defects in real materials are complicated. Nevertheless the atomistic calculations do indicate that the basic physics of fracture and flow can be understood on the basis of simple (but atomistic) mechanical principles. The transient nature of these nonequilibrium problems underscores the need for simpler steady-state simulations. To carry the steady state simulations out requires the new methods of nonequilibrium molecular dynamics that we discuss in Chapter IV. We return to the subject of plastic flow in Section H of that chapter.

Bibliography for Chapter III

A very strong case for the usefulness of pair potentials in describing the properties of the rare gases has been made by John A. Barker in "Pair and Many-Body Interactions in Dense Rare-Gas Systems", to appear in Physical Review Letters (1986).

Both diffusion and viscosity in two-body periodic systems have been considered from the point of view of the Boltzmann equation. For references to this work see W. G. Hoover, "Nonlinear Conductivity and Entropy in the Two-Body Boltzmann Gas", Journal of Statistical Physics **42**, 587 (1986) as well as Section G.4 of Chapter IV of these notes.

References to the vast literature on approximate numerical solutions of the Boltzmann equation as applied to aerodynamic flows can be found in K. Nanbu, "Interrelations Between Various Direct Simulation Methods for Solving the Boltzmann Equation", Journal of the Physical Society of Japan **52**, 3382 (1983). See also E. Meiburg, "Direct Simulation Techniques for the Boltzmann Equation", Deutsche Forschungs-und-Versuchsanstalt für Luft-und-Raumfahrt Report number FB 85-13, Göttingen (1985). This outstanding work should appear shortly in The Physics of Fluids.

Considerable atomistic work has been carried out on fracture and flow problems. See, in addition to Bill Moran's University of California at Davis/Livermore Ph. D. Thesis, "Crack Initiation and Propagation in a Two-Dimensional Triangular Lattice" (1983), the following references, all taken from the Physical Review and describing work supported by the Army Research Office involving the efforts of W. T. Ashurst, J. A. Blink, B. L. Holian, W. G. Hoover, A. J. C. Ladd, B. Moran, and G. K. Straub: "Shock-wave Structure *via* Nonequilibrium Molecular Dynamics and Navier-Stokes Continuum Mechanics", **22A**, 2798 (1980); "Fragmentation of Suddenly Heated Liquids", **32A**, 1027 (1985); "Microscopic Fracture Studies in the Two-Dimensional Triangular Lattice", **14B**, 1465 (1976); "Plastic Flow in Close-Packed Crystals *via* Nonequilibrium Molecular Dynamics", **28B**, 1756 (1983). See also the references to Dale Huckaby's defect free energy calculations cited at the end of Chapter II.

IV. NONEQUILIBRIUM MOLECULAR DYNAMICS

IV.A Motivation for Generalizing Newton's Equations of Motion

Nonequilibrium molecular dynamics is a modification of Newton's mechanics. About a dozen years ago simulations of viscous flows and heat flows led to the development of these nonequilibrium techniques. **Bill Ashurst's** 1974 thesis at the University of California at Davis, "Dense Fluid Shear Viscosity and Thermal Conductivity *via* Nonequilibrium Molecular Dynamics", is a pioneering example. In addition to Newton's "applied" forces F_a (interatomic, gravitational, electromagnetic, and the like) and "boundary" forces F_b, we consider systems with additional "constraint" and "driving" forces:

$$m\ddot{r} = F_a + F_b + F_c + F_d.$$

The extra forces can be chosen so as to study systems away from equilibrium. This does not necessarily mean that energy conservation cannot be used to check the calculations or that the simulations will not be reversible in a mathematical sense. Most of the simulations do carry over the deterministic constants of the motion, as well as the formal time reversibility, and the Lyapunov instability associated with Newtonian mechanics.

Why change Newtonian mechanics? The reasons are pragmatic. It is simpler and cheaper to simulate and study those processes we understand and to explore those we do not yet understand using modified forms of mechanics. Boundaries are always present in real physical systems. These cause difficulties in computer simulations. There is a tendency for particles to order at physical boundaries and to behave in an atypical way there. **Figure 1**, from Farid Abraham's recent review, shows a solid-fluid interface.

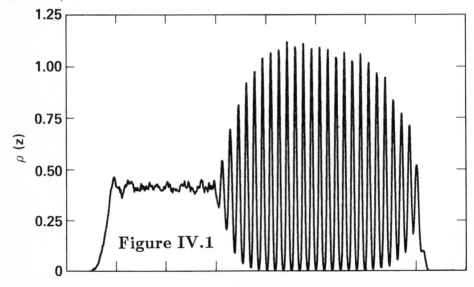

Figure IV.1

The mean density is plotted as a function of distance normal to the crystal boundary. Such density profiles suggest that the effective range of interfaces (within which atomistic simulations are necessary for quantitative work) is comparable to that found for cracks and dislocations, of order ten atomic diameters. To minimize the relative importance of such interfacial effects a large system has to be used. With periodic boundaries the interfaces present at system boundaries are eliminated. Thus bulk properties can be measured using *much* smaller systems. In nonequilibrium simulations it is not obvious how to introduce temperature and velocity gradients using periodic boundaries. A *trick* needs to be used to force a periodic system to support a heat flux or an anisotropic momentum flux. The constraint and driving forces introduced by nonequilibrium molecular dynamics make it possible to do this.

John Barker has emphasized that the extra constraint and driving forces associated with nonequilibrium molecular dynamics bear a close resemblance to the electromagnetic fields used in spectroscopy. In either case external fields are used to modify and to probe dynamical behavior. Without such constraint and driving techniques it is extremely difficult to study nonequilibrium viscous or heat conducting steady states. With the extra forces the systems studied will perhaps be simple enough to stimulate quantitative theoretical treatment. Example nonequilibrium problems which can be handled theoretically include the two-body diffusive and viscous flows, which can be treated by a simple modification of the Boltzmann Equation, and the more complicated three-body heat-flow problem. These applications, as well as many-body simulations, are the subject of this Chapter.

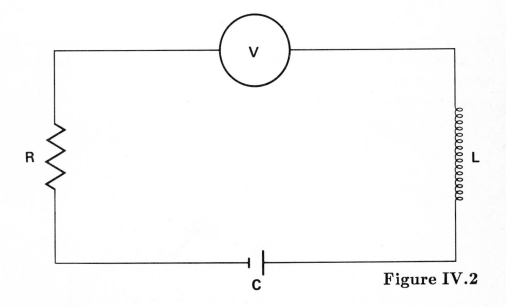

Figure IV.2

IV.B Control Theory and Feedback

"Control Theory" formalizes the connection between observed dependent variables and the specified independent variables controlled by the experimenter or simulator. Mechanical controls, operating with feedback based on the observed values of dependent variables, are relatively familiar. For instance, a centrifugal governor can be used to reduce the fuel supply if the rotation of a shaft becomes too rapid. Electronic controls are even simpler to design. The circuit shown in **Figure 2** combines inductive, resistive, and capacitive elements to produce a voltage V with differential, proportional, and integral responses to the current I:

$$V = L\dot{I} + RI + \int (I/C)\, dt,$$

where L, R, and C are the circuit's inductance, resistance, and capacitance, respectively.

If we wish to control a deviation Δ through a control variable ς it is natural to use the same procedure:

$$\varsigma = A\dot{\Delta} + B\Delta + C \int \Delta\, dt.$$

Such *linear* relationships have been much used because they can be solved analytically, using Laplace transforms, or electronically, using simple circuits. The coefficients can be chosen empirically, or in accord with a physical principle such as Gauss'. Simple *linear* control equations are adequate to describe *all* of the nonequilibrium molecular dynamics work carried out so far. This includes the simulation of diffusive, viscous, and heat flows together with problems involving combinations of these flows.

IV.C Examples of Control Theory: "Isothermal" Molecular Dynamics

We discussed the application of Gauss' Principle of Least Constraint to the simulation of systems at constant temperature in Section D of Chapter I. Gauss' Principle corresponds to "differential" control. The principle leads to additional constraint forces having the form of frictional forces:

$$\dot{p} = F_a + F_c = F_a - \varsigma p.$$

Exactly the same expression for the constraint forces would be the likely choice of a control engineer asked by a physicist to help design equations of motion with specified properties. The friction coefficient ς differs from the better-known friction coefficients occuring in the Stokes-drag friction on a body moving in a viscous fluid (proportional to the size of the body and to the fluid viscosity) or in the stochastic Langevin equation describing the interaction of a test particle with an irreversible heat bath. In both these cases the friction coefficient is a positive constant. In Stokes drag the moving particle is typically accelerated by a constant external force (gravity or an electromagnetic field) to offset the drag. In the Langevin approach to Brownian motion, the moving particle is instead accelerated by stochastic forces with random magnitude and orientation. In Gaussian isothermal dynamics Gauss' friction coefficient ς can be either negative or positive. It varies with time and has a time-averaged value of *zero* at equilibrium. If we consider ς as a control variable Gauss' Principle is an example of *differential control*. In the equilibrium case ς is proportional to the time rate of change of the potential energy.

$$\varsigma = -\dot{\Phi}/(2K),$$

where K is the kinetic energy.

Hermann Berendsen has suggested the use of *proportional control* to keep either the temperature or the pressure near its desired value. In Berendsen's schemes the control variable ς is again used to control the temperature, but ς is chosen to be proportional to the difference between the present and desired values of the temperature or pressure. In the latter case the volume can be changed by an amount proportional to the pressure error (defined as the difference between the virial theorem pressure and the desired pressure). These proportional-control equations of motion are not time-reversible and are, for this reason, harder to analyze theoretically. Irreversible motion equations like them have been studied for more than 100 years. Both *Rayleigh's equation*,

$$\ddot{x} = -x - \dot{x}\left(\dot{x}^2 - 1\right),$$

and its time derivative, *van der Pol's equation*,

$$\ddot{v} = -v - \dot{v}\left(3v^2 - 1\right),$$

are well-known examples. If we rewrite van der Pol's equation, replacing the dependent variable $\sqrt{3}\,v$ by x, we find again an example of proportional control, as applied to a harmonic oscillator:

$$\ddot{x} = -x - \dot{x}\left(x^2 - 1\right).$$

But now van der Pol's equation provides feedback to the friction coefficient on the basis of a potential, rather than a kinetic, energy. Solutions of the Rayleigh and van der Pol equations, for initial values of q and $\dot{q} = p$ both equal to 1, are shown in **Figure 3**. Both solutions exhibit typical (mathematically) irreversible behavior, converging exponentially fast to steady limit cycles.

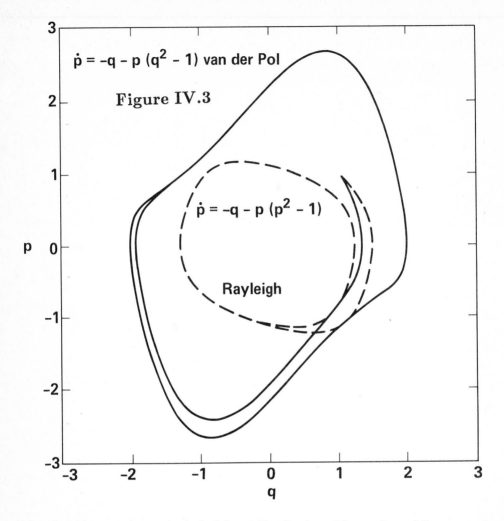

$$\dot{p} = -q - p\,(q^2 - 1) \text{ van der Pol}$$

Figure IV.3

$$\dot{p} = -q - p\,(p^2 - 1)$$

Rayleigh

Although *neither* equation can be derived from a Hamiltonian, *either* can be usefully expressed as a pair of first-order ordinary differential equations, as indicated in the **Figure**. Rayleigh's 1883 Philosophical Magazine paper "On Maintained Vibration" describes multitudes of applications for these simple feedback equations. Both Rayleigh's and van der Pol's equations are examples of *proportional control*.

Nosé introduced the use of *integral control* variables in statistical mechanics. His approach was described in Section E of Chapter I. Nosé suggested choosing ς proportional to the time integral of either the pressure or temperature deviation from its desired value. This procedure has the twin virtues of time reversibility and a direct connection to Gibbs' statistical equilibrium theory. The new equations generate the canonical and constant-pressure ensembles. It is straightforward to generalize Nosé's approach in two ways: first, to nonequilibrium systems; second, to systems in which higher moments of the velocity, or higher derivatives of the volume, are controlled. Nosé's equations are examples of *integral control*.

The distinction between differential and integral control is not entirely clearcut. In the limit that the response time of Nosé's *integral* control approaches zero, Gauss' *differential*-control equations of motion are recovered.

IV.D Heat Conducting Chain - 9 Examples

In Section D of Chapter II we discussed the number-dependence of a one-dimensional harmonic chain. The rest configuration of a three-atom chain is shown at the top of **Figure 4**. In the equilibrium case such a chain has no mechanism for scattering, and is relatively uninteresting. Away from equilibrium this can be changed, by using constraint and driving forces. An unperturbed harmonic chain can support either standing or travelling waves. The two lower drawings in **Figure 4** suggest two fundamental standing-wave vibrations of a three-particle chain. Such a chain is the simplest for which it is possible to consider the transfer of heat *via* phonons. From the standpoint of control theory the travelling waves (phonons), which have constant total energy, potential energy, and kinetic energy, are simpler than standing waves.

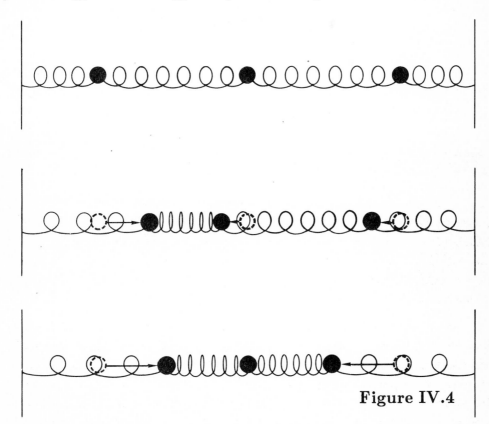

Figure IV.4

In a *standing wave* the displacements are products of space-dependent and time-dependent functions. The potential and kinetic energies vary sinusoidally in the time while the total energy is fixed. In a travelling wave, or phonon, the chain passes periodically through the lower two configurations indicated in **Figure 4**, but in such a way that the sums of the three particles' kinetic and potential energies remain constant. Thus phonons are already *exact solutions* of the control-theory equations which seek to control kinetic, potential, or total energy. We will investigate this three-body system in the case that an additional energy-sensitive driving force induces a preference for the right-moving phonon over the left-moving one.

For convenience we choose the rest positions of the three particles as follows:

$$x_1 = 1/2; \quad x_2 = 3/2; \quad x_3 = 5/2.$$

The motion of the right-moving and left-moving phonons are described by the equations

$$\delta x_1 = A\cos(\sqrt{3}\,t \pm 1\pi/3);$$

$$\delta x_2 = A\cos(\sqrt{3}\,t \pm 3\pi/3);$$

$$\delta x_3 = A\cos(\sqrt{3}\,t \pm 5\pi/3).$$

where plus signs correspond to the left-moving and minus signs to the right-moving phonon. We have chosen to set the mass and force constants all equal to unity, so that the frequency is $\sqrt{3}$. The wavelength is 3. A is the phonon amplitude. The phase space for this three-particle system is four-dimensional (three coordinates and three momenta, but with the total displacement and total momentum zero) so that it is useful to look at projections onto a two-dimensional subspace to visualize the motion.

It is convenient to plot the momentum for one particle as a function of the coordinate of one of its neighbors. If we choose $p_2(x_1)$ the right-moving phonon generates an ellipse, traversed counter clockwise, lying primarily in the first and third quadrants. The left-moving phonon, in this same space, corresponds to a counter clockwise orbit lying primarily in the second and fourth quadrants. We will use the $p_2(x_1)$ projections to diagnose the effect of *nine* different types of Non-Newtonian constraint forces, F_c. In addition to the *constraint* force F_c, which plays the rôle of a thermostat, we need a *driving* force F_d appropriate to heat flow. Such forces must be chosen to be consistent with the results of Green-Kubo linear response theory. Evans and Gillan showed that driving forces proportional to each particle's contribution to the energy and to the potential part of the pressure P^ϕ should be used:

$$\dot{p} = F_a + F_d = F_a + \lambda\left(\delta E + V\,\delta P^\phi\right).$$

The rate at which work is done by these driving forces, which is the "Power Loss", is related to the resulting heat current through the heat conductivity in the usual way:

$$\text{Power Loss} = T\dot{S} = Q^2 V/(\kappa T).$$

This relationship will be established in Section I of this Chapter. It follows easily from the Heat Theorem applied to a bar maintained at a higher temperature $T_H \sim T$ at one end and a lower temperature $T_C \sim T$ at the other, resulting in a steady heat flux Q.

To apply the Evans-Gillan heat-current driving force to a one-dimensional chain requires only a way of specifying each particle's energy and pressure contributions. For Particle 2 these are

$$E_2 = (p_2^2 + \phi_{12} + \phi_{23})/2;$$

$$V P_2^\phi = \left[(1 + \delta x_2 - \delta x_1)(\delta x_1 - \delta x_2) + (1 + \delta x_3 - \delta x_2)(\delta x_2 - \delta x_3) \right]/2.$$

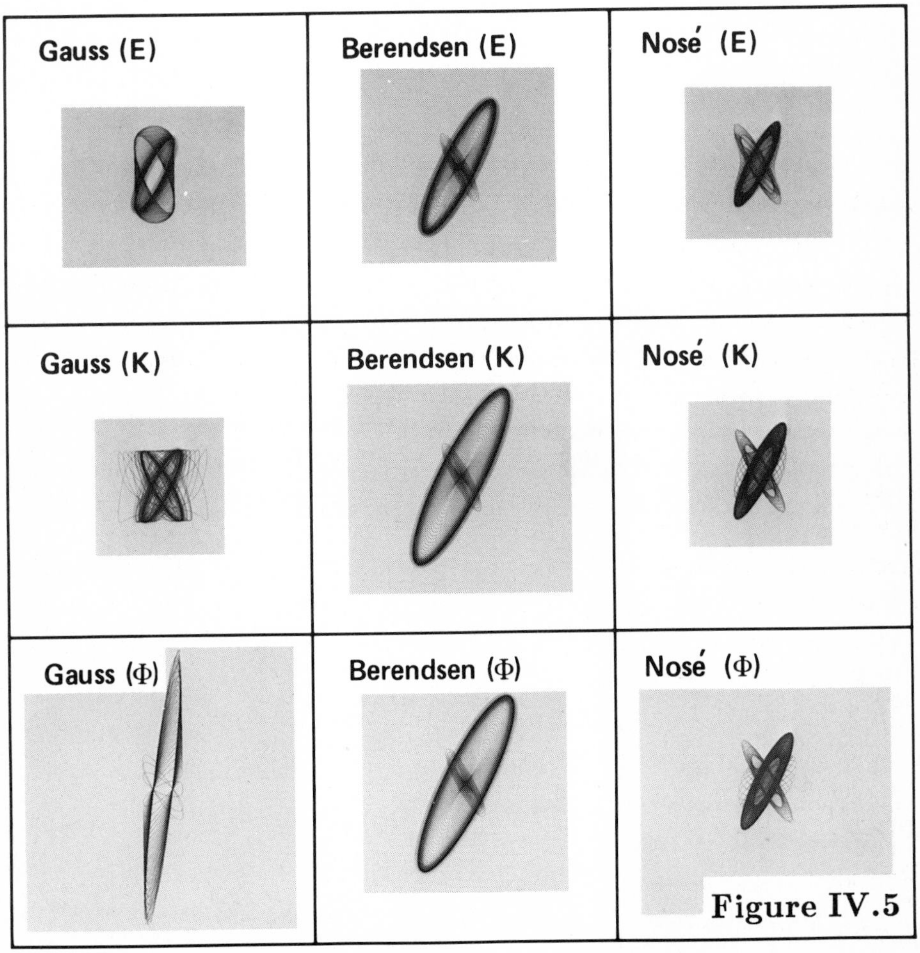

Figure IV.5

To compensate for the power absorbed by the chain we can use constraint forces designed to control any one of the three energies: total (E), kinetic (K), or potential (Φ). Each of these three energies could be controlled by any of the three kinds of constraint forces: differential (Gauss), proportional (Berendsen), or integral (Nosé). Trying out all nine combinations shows that, qualitatively, eight of the nine actually work, leading to similar steady solutions. All nine calculations are shown in **Figure 5**. The *unstable* solution shown in the lower lefthand corner corresponds to keeping the potential energy of the system fixed. This approach is a clear failure. The chain simply traverses the same constant-potential set of energy states at faster and faster rates with the kinetic energy diverging as time goes on. In the eight steady solutions, and with moderate field strengths, all of the energy eventually contributes to the right-moving phonon, which maximizes the heat flux at fixed energy. At higher fields, the differential (Gauss) and integral (Nosé) reversible controls are both more stable than Berendsen's irreversible proportional control. From the *qualitative* standpoint, in eight of the steady state solutions the current reaches the *same* dynamical state, a phonon carrying heat current to the right. But the amplitude varies with the particular method chosen. From the *quantitative* standpoint the eight steady solutions differ considerably from one another. The amplitudes of the eight stable motions shown in the **Figure** vary by more than a factor of two, with correspondingly large differences in the resulting conductivities.

For a one-dimensional chain the zero-field conductivity diverges because reducing the field strength λ does not reduce the steady-state current, only the rate at which the system achieves it. This lack of a limiting conductivity stems from the absence of an efficient scattering mechanism in one dimension. Thus the conductivity κ varies inversely with the field in one dimension. In two dimensions, three hard disks are already sufficiently complex to have a more complicated dependence of current on field strength. For the disks, the conductivity *increases gradually* as λ is made smaller, varying logarithmically instead of inversely with current for the smallest λ's that have so far been investigated. This logarithmic dependence is consistent with the predictions of a complex theoretical approach called "mode-coupling theory", but because that theory is based on the interaction of sound waves, I believe that a simpler approach will eventually be developed.

IV.E Linear and Nonlinear Response Theory

Conventional linear response theory is based on adding a perturbation δH to the Hamiltonian and using the resulting changed distribution function

$$f/f_o = e^{+\delta H/(kT)}$$

to compute averages. In the linear theory these averages involve the equilibrium distribution function, f_o. The theory can be used to generate nonlinear nonequilibrium dynamical equations which produce limiting fluxes consistent with the linear theory.

As an illustration consider heat conduction. From irreversible thermodynamics, the dissipation $T\dot{S}$ necessary to maintain a heat current Q between hot and cold reservoirs at temperatures T_H and T_C is $Q^2V/(\kappa T)$. The generalization of the one-dimensional Evans-Gillan external driving force to three dimensions, which generates a heat current in the x direction Q_x, and produces exactly the *same* dissipation, with T_H and T_C again replaced by T, has driving-force components in all three spatial directions:

$$F_d = \lambda \left[\delta E + V \delta P_{xx}^\phi, V \delta P_{xy}^\phi, V \delta P_{xz}^\phi \right].$$

Thus the distribution function at time t, $f_o e^{+\Delta E/(kT)}$, where ΔE is the work done by the driving force during the time interval from 0 to time t, is

$$f = f_o e^{[\lambda V/(kT)] \int_o^t Q_x \, ds}.$$

Using the linear expansion of this exponential, the average long-time nonequilibrium current that results as t approaches ∞, $\iint f Q_x(t) \, dq \, dp$ can be expressed as an equilibrium autocorrelation integral:

$$\langle Q_x V \rangle_{ne} = \left[\lambda V^2/(kT) \right] \int_o^\infty \langle Q_x(0) Q_x(t) \rangle_{eq} \, dt.$$

Jaynes and Zubarev suggested that not just linear distribution functions, but also *nonlinear* distribution functions can be obtained in this way. For heat conductivity, the distribution function is just that given above, without expanding the exponential. Substituting this distribution into the appropriate time evolution equation for f demonstrates that f is indeed a solution of the nonequilibrium equations of motion. But the solution has serious flaws. *It doesn't apply to systems with periodic boundaries and it cannot be used to describe steady states.* We will point out how the solution must be modified in the next section.

IV.F Diffusion and Viscosity for Two Hard Disks

The simplest nonlinear transport problem is the mutual diffusion (caused by an external field E, *with units of force*) of two isothermal disks. With periodic boundaries this problem corresponds to an infinitely high and infinitely wide "Galton Board", in which one of the disks "falls" horizontally through a triangular lattice of "pins" (periodic images of the second disk). Finite Galton Boards, such as the fifteen-pin example shown in **Figure 6**, are used—at least in "thought experiments"—to generate binomial distributions. If an ensemble of falling particles were "dropped", from the left, into a board such as that shown in the **Figure**, making inelastic collisions at each column of pins, we would expect to find a binomial distribution at the righthand boundary. In the infinite-board limit the binomial distribution becomes a Gaussian. Here we study a *periodic* Galton board. One periodic parallelogram cell is indicated in **Figure 6**.

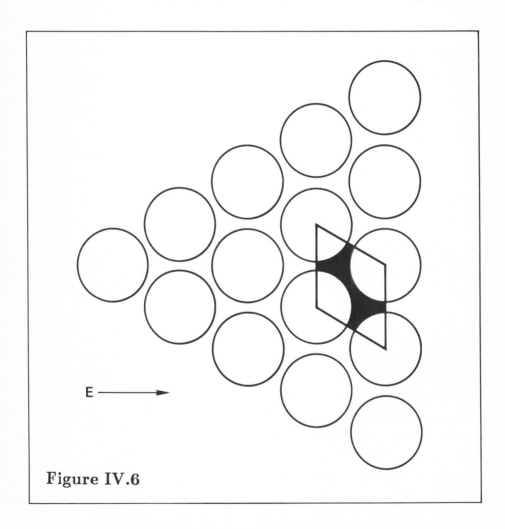

E ⟶

Figure IV.6

In a real Galton Board, inelastic collisions would eventually produce a steady-state distribution of density within the cell, the deviation from uniform density depending upon the strength of the accelerating gravitational field. The problem is easier to treat if a constraint force is used to maintain the falling particle's temperature. If the field, *with units of force*, is of strength E and the kinetic energy is kept constant by using an isokinetic friction coefficient ς, the exact solution of Boltzmann's time-dependent equation, in the relaxation-time approximation, can be written down directly, by summing up the distribution of particles which have not collided and the distribution of particles which have collided one or more times.

To start out, consider the equations of motion for the field-driven diffusion problem with the mass and speed and diameter of both disks set equal to unity. The equations are simplest if the field is oriented parallel to the x axis, as shown in **Figure 6**. It is convenient to choose $p^2 = 1$ so that the momenta can be described with polar coordinates with a radius of unity. The resulting equations of motion, with a plus sign for Particle 1 and a minus sign for Particle 2, are then as follows:

$$\dot{x} = p_x = \cos(\theta);$$

$$\dot{y} = p_y = \sin(\theta);$$

$$\dot{p}_x = F_x - \varsigma p_x \pm E;$$

$$\dot{p}_y = F_y - \varsigma p_y.$$

The *same* equations result if the motion is described in a frame fixed on either of the two particles. But then the *reduced* mass, $1/2$ in the case that both particles have unit mass, must be used and the speed of the moving particle must be doubled. Between collisions F vanishes, and the equations can be simplified:

$$\dot{p}_x = -\dot{\theta}\sin(\theta) = -\varsigma\cos(\theta) \pm E; \quad \dot{p}_y = +\dot{\theta}\cos(\theta) = -\varsigma\sin(\theta).$$

The requirement that the friction coefficient ς be chosen to make the sum $p_x\dot{p}_x + p_y\dot{p}_y$ vanish gives

$$\varsigma = E\cos(\theta); \quad \dot{\theta} = -E\sin(\theta).$$

The steady-state relaxation-time Boltzmann equation, in polar coordinates, has a relatively simple form:

$$\partial(f\dot{\theta})/\partial\theta = (df/dt)_{collisions} = (f_o - f)/\tau,$$

where τ is the mean time between collisions.

To solve this equation at any time t we need only notice that $f(\theta)$ can always be written as a sum formed by adding the distribution of those particles streaming from θ_o to θ during the time t which haven't collided at all, $f_{initial}\, e^{-t/\tau}[\sin(\theta_o)/\sin(\theta)]$, to all particles which had their most recent collision a time s in the past. Because the collision rate is independent of f, the latter term has the same exponential form, $f_o\, e^{-s/\tau}$. The ratio of sines corresponds to the Jacobian of the transformation from time 0 to time t.

$$d\theta_o/d\theta = \dot\theta_o/\dot\theta = \sin(\theta_o)/\sin(\theta).$$

It is necessary to know the change in angle since the last collision. From the equation of motion the time t required to stream from θ_o to θ is

$$t = (\tau/E)\ln\left[\tan(\theta_o/2)/\tan(\theta/2)\right].$$

Making this substitution gives the solution of the relaxation-time Boltzmann Equation

$$f/f_o = (f_{initial}/f_o)\, e^{-t/\tau}\left[\sin(\theta_o)/\sin(\theta)\right] + \int_\theta^{\theta_o} e^{-s/\tau} d\beta/(E\sin\theta),$$

where the integral includes all angles $\theta_o \geq \beta \geq \theta$ from which θ can be reached during the time t and where s is the time required to stream from β to θ. Direct substitution into the Boltzmann equation verifies this solution. The resulting *nonlinear* dependence of the diffusion coefficient on the field strength is shown in **Figure 7**.

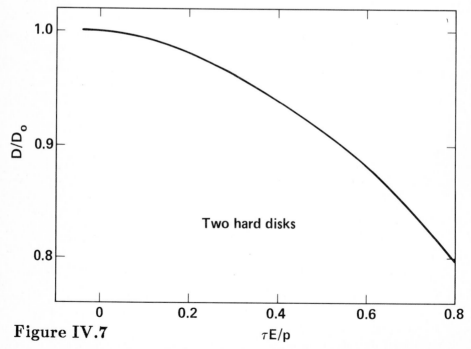

Figure IV.7

This two-body Boltzmann Equation solution is useful in illustrating a pitfall associated with Jaynes' information theory. That theory suggests that the exact nonequilibrium distribution function can be found, at any time t, by specifying all relevant properties of the system in question at that time and, simultaneously, maximizing the entropy, $-k \langle \ln f \rangle$. **Figure 8** shows the solution of the Boltzmann equation for two disks as well as the information theory solution. The angle θ measures the direction of Particle 1's velocity relative to the field direction. Information theory takes the point of view that the distribution function f can be found by maximizing the entropy subject to all known constraints. What are the constraints? If one matches the density, temperature, and current from the Boltzmann equation the resulting information-theory distribution, marked "Information Theory" in **Figure 8**, is very different from that of the Boltzmann equation, marked "Exact". The field corresponding to the case shown in the **Figure** is $1/(2\tau)$.

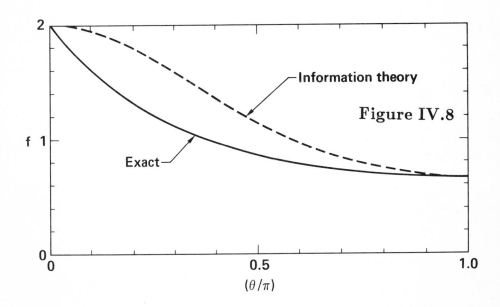

Why is the information-theory distribution wrong? The difficulty is that the information-theory distribution is not a *steady* solution of the equations of motion. The steadiness requirement has to be included in order for the information-theory approach to make correct predictions. That is, not only must the current have a specified value. Also, the first, second, third, ... time derivatives of the current must *all* vanish. Because implementing this infinite string of restrictions is fully as difficult as solving the Boltzmann Equation in the first place, it is not clear that the theory is a useful one for kinetic-theory problems far from equilibrium. This is unfortunate because there are no other theories at present which appear to be useful in solving these problems.

The exact solution of the two-dimensional diffusion problem can be generalized to apply to the three-dimensional case of hard spheres diffusing in an external field as well as to the two- and three-dimensional cases of viscous flow. In the viscous-flow case the equations of motion include the strain rate, $\dot{\epsilon} = du_x/dy$, and again a friction coefficient ς. The viscosity which results for the corresponding two-dimensional hard-disk case is shown in **Figure 9**. The disk diameter is σ. As the strain rate increases beyond the inverse collision time the viscosity drops to a small fraction of the low strain rate limit calculated from the analytic solution of the Boltzmann equation. This drop in viscosity, called "shear thinning", is also typical of results obtained with many particles, not just two.

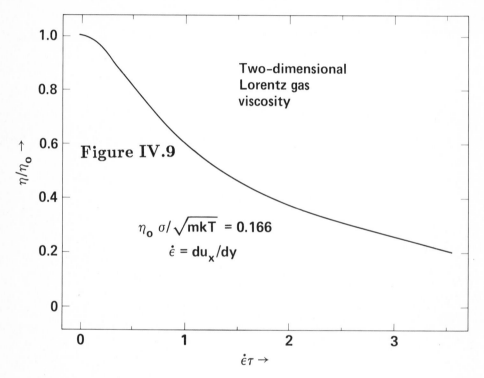

IV.G Simulation of Diffusive Flows

IV.G.1 Fractals.

Diffusion generates particle trajectories which are "fractal objects" with interesting geometric properties. We begin by describing these objects. A fractal object generally has a dimensionality less than that of the space in which it is embedded. The fractal dimensionality is usually not an integer. The main idea is that the "volume" of the fractal object varies systematically with measurement *scale* . Thus a fractal object remains irregular, with a nondifferentiable boundary, at an arbitrarily fine scale. The idealized mathematical concept can be a physically useful one whenever such scaling can be applied over length scales spanning a decade or more.

If, for instance, we estimate the "volume" of an ordinary one-, two-, or three-dimensional object in three-dimensional space by multiplying by δ^3 the number of cells of sidelength δ intersecting a part of the object, as a function of the measuring length δ, we can see that a line will have a limiting "volume" varying as δ^2, an area will have a "volume" varying as δ^1, while a "real" volume approaches a definite limit as δ approaches zero. Thus a natural extension of the usual dimensionality is just $D_{fractal} = 3 - (d \ln \text{"}volume\text{"}/d \ln \delta)$. This new fractal dimension closely resembles the "Hausdorf dimension".

The usual textbook example of a fractal object is a Cantor set. This is a line segment on which the following operation has been repeated an infinite number of times: delete the middle third of the line segment, thereby generating two new shorter segments. With each repetition of the process the line segments which were formed in the previous deletion have their middle thirds removed. If this process is repeated N times then the actual remaining length of the line is $(2/3)^N$ times the original length. Thus changing the measuring length δ by a factor of $1/3$ changes the apparent line length by a factor of $2/3$. The fractal dimension of the fractal object formed in the limit is

$$D_{fractal} = 1 - \left[\ln(2/3)/\ln(1/3) \right] = 0.631.$$

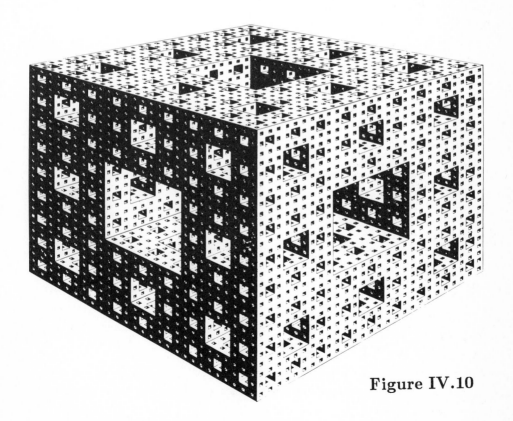

Figure IV.10

Figure 10, reproduced from Benoit Mandelbrot's stimulating book, shows an extension of this Cantor-set idea to three dimensions. If three intersecting square holes, of width 1/3, are bored in a cube of unit sidelength 20/27 of the original volume remains. Repeated iterations of this operation result in the "Sierpinski Sponge". The dimensionality of the sponge is $D_{fractal} = 3 - \left[\ln(20/27)/\ln(1/3)\right] = 2.727$.

The trajectory of a particle diffusing in three-dimensional space according to Fick's law, with $\langle x^2 \rangle = \langle y^2 \rangle = \langle z^2 \rangle = 2Dt$, can also be analyzed as a fractal object with a dimensionality less than that of the three-dimensional space in which it is embedded. If we follow a diffusing particle for N randomly oriented jumps of length λ then the distance traveled along the trajectory, $N\lambda$, takes the particle only a distance of order $\sqrt{N}\,\lambda$ from the origin. This result follows from the Central Limit Theorem or from Fick's Laws of diffusion, given below. This same kinetic-theory random-walk Brownian-motion result would obtain if the moving particle made only $N/(n^2)$ longer jumps of length $n\lambda$, where n is any positive integer. The total length of the path, using the longer jumps would be smaller, just $N\lambda/n$. The "longer jumps", of length $n\lambda$, can play the rôle of the measuring length δ introduced above. The volume associated with this description of a Brownian path would be $(N\lambda/n)\,(n\lambda)\,(n\lambda) = nN\lambda^3$. Thus the apparent "volume" of a Brownian-motion trajectory varies linearly with the measuring length and is therefore *two*-dimensional from the fractal point of view. It is clear that the fractal dimensionality is a far-from-complete description of the geometry. Both a smooth two-dimensional surface and a Brownian trajectory in three dimensional space have the *same* fractal dimensionality, but they are very different from the geometric point of view.

IV.G.2 Fick's Laws

Macroscopic diffusion is conventionally described by Fick's first and second laws. The fundamental first-law relationship is a *linear* one, stating that the current responds to the gradient in concentration in a linear way:

$$J = -D\,\nabla\rho, \qquad \text{(Fick's first law)}$$

where D is the diffusion coefficient. From the microscopic viewpoint, in which each particle has its own velocity, an equilibrium fluid contains two fast and powerful currents, equal in magnitude and opposite in direction. Because these currents are clearly proportional to the density it is reasonable that a density gradient would lead to a resultant current proportional to that gradient. If D is constant then the rate at which concentration builds up in a volume element, from the divergence of the current J, can be expressed in terms of the second derivative of ρ:

$$(\partial\rho/\partial t) = D\,\nabla^2\rho. \qquad \text{(Fick's second law)}$$

A useful solution of this equation describes the one-dimensional spreading of a spatial Gaussian delta function with time:

$$\rho(t) = \rho_o\, e^{-x^2/4Dt}/(4\pi Dt)^{1/2}.$$

Here D is the diffusion coefficient, *not* the dimensionality. This solution indicates that the width of the (Gaussian) distribution, shown in **Figure 11**, increases as the square root of the time and that the amplitude varies as the inverse square root. If the diffusion occurs in three-dimensional space then normalization requires that the amplitude vary as $t^{-3/2}$.

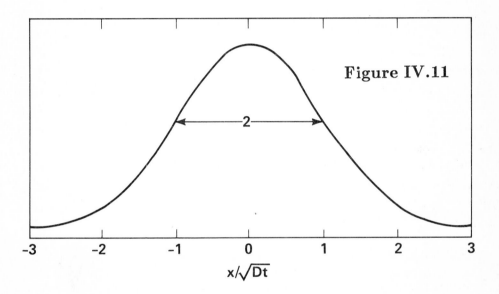

Figure IV.11

How large is the diffusion coefficient D? According to kinetic theory D is of order λv, where λ is the mean free path and v is the root-mean-square speed. For a gas this product is about one square centimeter per second. For a liquid the mean free path is smaller, perhaps 0.1 Ångstrom rather than 5000 Ångstroms. Thus the liquid diffusion coefficient is smaller, about 10^{-5} centimeters^2/second. The Gaussian solution following from Fick's Second law also predicts that the higher moments can be expressed in terms of the lower ones. In a Gaussian distribution *all* the moments are interrelated. For instance,

$$\langle x^4 \rangle = 3 \langle x^2 \rangle^2.$$

But the evidence from low-density kinetic theory, supported by relatively low-density molecular dynamics simulations, is that this relation fails for diffusion, so that Fick's laws apply only at length scales exceeding a few free paths.

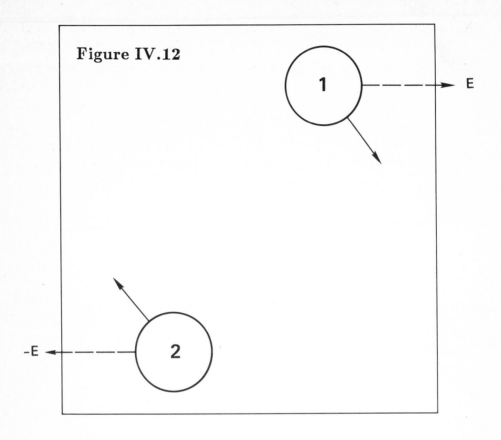

Figure IV.12

IV.G.3 Irreversibility and Heating

Diffusion can be measured in two ways, by observing the spreading out of equilibrium concentration fluctuations, or by measuring the steady current which results when an applied field accelerates particles. In either case the current arises so as to dissipate energy stored in the form of chemical potential. We can most easily understand the dissipation associated with diffusion by considering the case in which the driving force is an external field.

For definiteness, consider two hard spheres again, with mass m and diameter σ, as shown in **Figure 12**, with periodic boundaries and a field $+E$, *with units of force*, pushing Sphere 1 to the right, with a field $-E$ pushing Sphere 2 to the left. The velocities of the two spheres are indicated by solid arrows in the **Figure**. On the average the isotropic scattering of spheres in a center-of-mass frame guarantees that the velocities of an ensemble of pairs of colliding spheres, with initial velocities $+v$ and $-v$, are randomly oriented after one collision. If we calculate the ensemble-averaged velocity gained from the field by Sphere 1, the value is just $E\tau/m$, where τ is the mean time between collisions. The conservative field in which the particles move thus increases the kinetic energy at the rate $E^2\tau/m$.

Why doesn't the atmosphere, which is similarly composed of particles falling toward the earth under the influence of a nearly-constant gravitational field, heat up in a similar way? A consideration of the steady-state Boltzmann equation for an isothermal equilibrium atmosphere,

$$(\partial f/\partial t) + \dot{z}\,(\partial f/\partial z) - mg\,(\partial f/\partial p_z) = (\partial f/\partial t)_{collisions} = 0;$$

with f given by the Maxwell-Boltzmann equilibrium distribution, shows that both time derivatives vanish and that the other two derivatives cancel, each giving $\pm mgv_z[f/(kT)]$. Thus the velocity gained from the field exactly offsets the differential flow due to the equilibrium density gradient.

IV.G.4 Relaxation-Time Boltzmann Equation for Diffusion

Consider the homogeneous diffusion, under the influence of an external field, of a low-density gas. We will forgo the unforgettable experience of studying the rigorous solution of the Boltzmann Equation, as detailed in Chapman and Cowling's text. Instead we consider a much simpler, but fairly faithful, approximation to the complete theory, the "relaxation-time approximation". In the time-independent, homogeneous case the relaxation-time approximation to the Boltzmann equation,

$$(E/m)\,(\partial f/\partial v_x) = (f_o - f)/\tau,$$

can be solved as a power series in τ, the collision time. This establishes that the first-order perturbation to f_o, is given by the equation

$$f_1/f_o = (E/m)\big[mv_x/(kT)\big]; \quad f = f_o + f_1\tau + \cdots .$$

The corresponding distribution function is sketched in **Figure 13** in terms of the angle θ defining the velocity of Particle 1 relative to the field direction. This distribution leads to a current

$$J = (Nm\langle v_x\rangle/V) = (Nm/V)\big[E\tau/(kT)\big]\langle v_x^2\rangle_{eq} = NE\tau/V$$

linear in the field and giving again for the current

$$J = E\rho\tau/m.$$

The corresponding conductivity, $\rho\tau/m$, is obtained on dividing the current by the field strength. This formula should be thought of as a useful approximation, in the event that the collision time τ can be estimated. Alternatively this simple approach can be used to *estimate* collision times from *measured* conductivities. In the event that better than factor-of-two accuracy is required a many-body simulation can be carried out, as described below.

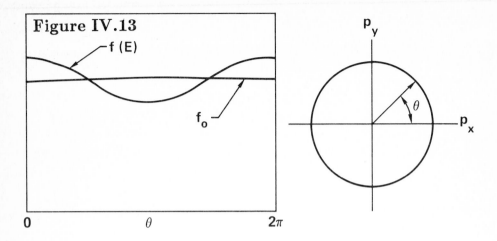

Figure IV.13

IV.G.5 Simulations

The diffusion coefficient has been the most-studied of all the transport coefficients, ever since the work of Berni Alder and Tom Wainwright on hard spheres. It is simplest to calculate because mass, unlike stress and energy, is a one-body property. Despite considerable work on the linear diffusion coefficient, appropriate to the zero-flux limit, nonlinear diffusion is not at all well understood. We saw in Section F that the two-disk results from the Boltzmann equation indicate a diffusion coefficient that *falls* with increasing field strength, just as viscosity falls with increasing strain rate. On the other hand simulations of larger systems of Lennard-Jones particles near the triple point show an *increase* in diffusivity with field strength. The liquid calculations so far carried out make it possible to compare diffusion coefficients calculated using three different methods.

(i) Green-Kubo evaluation of the velocity autocorrelation function.

(ii) Isothermal molecular dynamics in a constant field.

(iii) Isothermal molecular dynamics with a constant current.

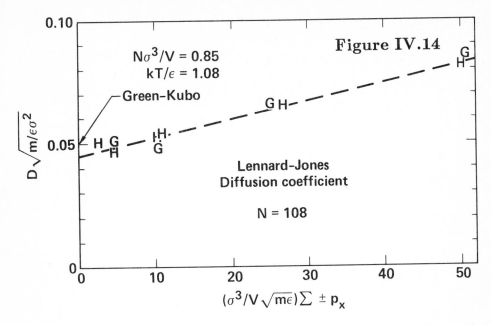

Figure 14 shows that the results from all three methods are internally consistent with one another. The Green-Kubo evaluation of the diffusion coefficient, which gives only the zero-current conductivity, has been in use ever since the pioneering calculations of Alder and Wainwright more than 30 years ago. Both "isothermal" methods, labelled H and G in the **Figure**, were implemented by constraining the kinetic energy contributions normal to the field direction, $\sum (p_y^2 + p_z^2)/(2m)$. The results indicated by H used a fluctuating current induced by adding an additional constant force in the x direction. The results indicated by G in **Figure 14** used instead a Gaussian constraint force to fix the current in the x direction. In the last calculation Gauss' Principle was used to find the forces necessary to keep the current constant. All three approaches predict the same diffusion coefficent (or conductivity) at zero field strength. The last two nonequilibrium methods also give nearly the same *nonlinear* conductivity, within the few percent uncertainties of the calculations. Other schemes, based for instance on keeping the energy, as opposed to the temperature, fixed, could also be used.

IV.H Simulation of Viscous and Plastic Flows

IV.H.1 Definitions and Typical Flows.

The three terms "viscosity", "plasticity", and "elasticity" are used to describe the dependence of shear stress on strain and strain rate. The shear stress (minus the appropriate pressure-tensor element) in fluids, caused by a strain rate, $\dot{\epsilon} = du_x/dy$ in the simplest case, is called *viscous* stress. Its magnitude is proportional to the shear viscosity η. The shear stress in flowing solids is described by the *plastic* yield strength Y. The shear stress in elastically deforming solids is described by the *elastic* shear modulus G or η. Molecular dynamics can treat all of these three idealized cases with the *same* microscopic equations of motion. If the shear stress is studied for a gradually deforming solid, the initial linear increase in stress with strain is the product of the elastic shear modulus G and the strain ϵ. If, for larger strains, as the solid deforms inelastically, the stress is found to have a steady value, Y, this shear stress is a "plastic yield strength" or "flow stress". If the shear stress for a deforming fluid is studied, this steady stress, divided by the strain rate $\dot{\epsilon}$, is a "viscosity coefficient", proportional to the shear viscosity η.

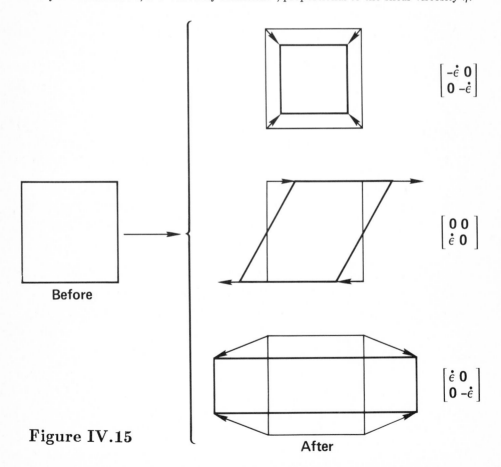

Figure IV.15

The three simplest flows are shown in **Figure 15** for a two-dimensional square volume element. From top to bottom these examples are compression, plane Couette flow, and simple irrotational shear. The three flows can be described in terms of the fluid velocity gradient or the solid displacement gradient, ∇u. In either case, fluid or solid, these tensors have the same form. The tensors corresponding to the three flow fields are shown in the **Figure**, using $\dot{\epsilon}$ to indicate the size of the nonzero elements in the tensor ∇u. In fluids the shear stress is described by the shear viscosity and the extra compressive stress by the bulk viscosity. This extra stress is not usually discussed for solids, and is thought, or hoped, to be relatively small compared to the yield strength, the solid-phase analog of viscosity.

The linear relation between stress and strain, or strain-rate, is only a useful approximation, but it is hard to go beyond that approximation. In the next order volume-element rotation must be taken explicitly into account, and there is no consistent way to do this using continuum mechanics. This subject generates considerable adrenalin in encounters between adherents to the *rigor mortis* school of continuum mechanics. Of course there is considerable data available, both from laboratory experiments and from computer experiments, describing the nonlinear response of fluids and solids to shear stress. In fluids it is typically found, in the second case above, plane Couette flow, that the viscosity is a decreasing function of strain rate.

The phenomenological description of stress in an ideal spatially isotropic elastic solid or an ideal viscous fluid is the same, with two parameters λ and η describing the resistance to changes in volume and shape. For elastic solids these are the Lamé constants. For viscous fluids these are the viscosity coefficients. In either case a linear combination of the two coefficients [$\lambda + \eta$ in two dimensions and $\lambda + (2/3)\eta$ in three dimensions] describes the dependence of stress on volume, with the coefficient η describing the dependence of stress on shape. The stress tensors have the form:

$$\sigma = \left[\sigma_{eq} + \lambda \nabla \cdot u\right] I + \eta \left[\nabla u + \nabla u^{transpose}\right].$$

For an elastic solid u is the vector displacement from an unstrained reference configuration. If the symmetrized tensor $\left[\nabla u + \nabla u^{transpose}\right]$ is nonzero then stress exists in the solid. For a viscous fluid the vector u is the stream velocity. If the symmetrized velocity gradient, called the strain-rate tensor $\left[\nabla u + \nabla u^{transpose}\right]$ is nonzero then the fluid has viscous stresses. The fluid rotation rate is described by the nonsymmetric part of the velocity gradient $\left[\nabla u - \nabla u^{transpose}\right]$, the "vorticity tensor". The fact that both the elastic solid and the viscous fluid stress tensors have the same form can be exploited in a limited class of problems for which σ is maintained by external forces (a sphere moving through a viscous fluid is an example.)

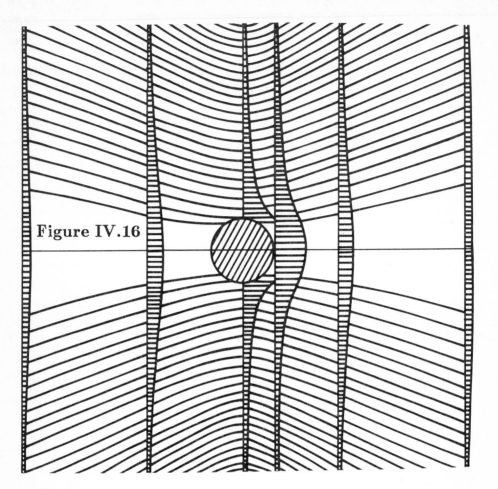

Figure IV.16

The flow around such a moving sphere is shown in **Figure 16**. Both velocity profiles (line segments proportional to the horizontal velocity) and streamlines (paths followed by elements of the fluid) are shown. This solution of the steady viscous flow problem corresponds exactly to the solution of an elastic problem in which an embedded rigid sphere is displaced relative to its rest position in an elastic medium. In that case the velocity profiles shown in the **Figure** correspond to displacement profiles. Even the two solutions of the fluid problem corresponding to sticking or slipping boundary conditions at the surface of the sphere have elastic analogs, depending upon whether or not the embedded sphere is linked to the continuum over its entire surface.

Viscosity, either shear or bulk, can be related to the lost work involved in carrying out a cyclic deformation. Suppose, for instance, that the magnitude of the driving strain varies sinusoidally in time with amplitude ϵ_o and frequency ω :

$$\epsilon = \epsilon_o \sin(\omega t); \quad \dot{\epsilon} = \epsilon_o \omega \cos(\omega t); \quad \sigma = \eta \dot{\epsilon}.$$

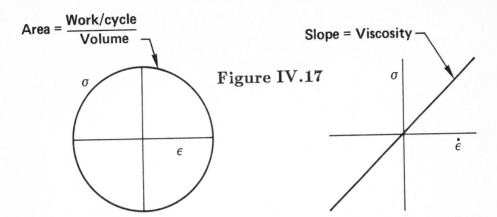

Figure IV.17

Then the work done per unit volume in the corresponding cycle shown in **Figure 17**, the integral of stress times strain rate, $\int \sigma \dot{\epsilon} dt$, is the area of the stress-strain ellipse:

$$(2\pi/\omega)\, \eta\, (\omega\epsilon_o)^2/2.$$

In order for this loss to be comparable to the thermal energy density, NkT/V, a frequency ω of order 10^{13} hertz is required. Thus, except in shockwaves, viscous heating is a relatively slow and insignificant process.

Viscous flows in fluids can be described by a diffusion equation having the same form as Fick's Second Law. In the simplest case, plane Couette flow, with $u_x = \dot{\epsilon}y$, the x component of the *linearized* equation of motion for a Newtonian viscous fluid is

$$(\partial u_x/\partial t) = (du_x/dt) = (1/\rho)\,(\partial \sigma_{yx}/\partial y) = (\eta/\rho)\,(\partial^2 u_x/\partial y^2),$$

so that the combination η/ρ plays the role of a diffusion coefficient. This combination of shear viscosity and density is called the *kinematic* viscosity.

We can see the importance of kinematic viscosity to a description of fluid flow by calculating the ratio of the time required for the shear motion of the $L \times L$ square volume element shown in **Figure 18** to diffuse away, L^2/ν, to the time required for the element to deform (to a shear strain of unity) L/v. The dimensionless ratio of these times is the "Reynolds Number" Re

$$Re = Lv/\nu.$$

For Reynolds numbers which are not too large, viscosity dominates and the flow is regular "laminar flow". For flows which are *not* dominated by viscosity (Reynolds numbers of a few thousand and above) the flow is "turbulent". It is interesting to see that a "thermodynamic limit" doesn't exist for flow problems. With a fixed velocity on the boundary, or a fixed velocity gradient, the large-system limit *always* becomes turbulent. The length scale at which this happens is called the Kolmogorov length. It is of order one millimeter for both the atmosphere and the ocean.

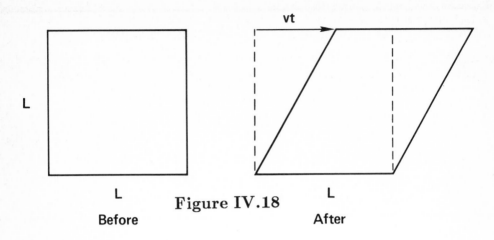

Figure IV.18

Before   After

IV.H.2 Andrade's and Enskog's Viscosity Models.

Andrade suggested the Einstein-like vibrational model shown in **Figure 19** for estimating the viscosity of a fluid. Imagine that fluid atoms oscillate (at a frequency $\nu_{Einstein}$ of order 10^{12} hertz) and that the fluid has a macroscopic velocity gradient $du_x/dy = \dot{\epsilon}$. Then, as an atom

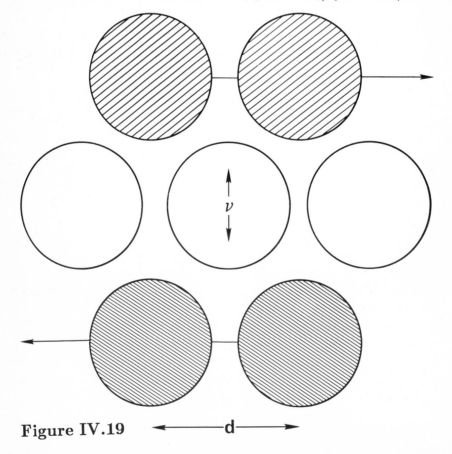

Figure IV.19

vibrates, it can transmit momentum between its neighbors. During a vibration time $1/\nu_{Einstein}$, these vibrations transmit momentum $2md \times \dot{\epsilon}$ through an area d^2, where d is the interparticle spacing and $\dot{\epsilon} = du_x/dy$ is the shear strain rate. Because this momentum flux, $-2\nu_{Einstein}m\dot{\epsilon}/d$, is *minus* the shear stress σ, the coefficient of proportionality $\sigma/\dot{\epsilon}$ is the viscosity coefficient

$$\eta_{Andrade} = 2\nu_{Einstein}m/d.$$

Numerical estimates (with $\nu_{Einstein} = 5 \times 10^{12}$ hertz, $m = 3 \times 10^{-23}$ grams, and $d = 3 \times 10^{-8}$ centimeters) give a viscosity of about 0.01 poise (the centimeter-gram-second unit of impulse or viscosity, pressure times time), about what is observed for simple fluids, such as water.

Enskog's model for fluid viscosity is a little more complicated. He estimates the enhanced transport of momentum through the mechanism of interparticle forces, using a hard-sphere model, and finds an increase of up to about a factor of twenty over the low-density Boltzmann equation value we estimate in the next section. This increase is reasonable in view of the fact that each collision, in a dense fluid, transports momentum a distance on the order of 10 free paths.

IV.H.3 Boltzmann Equation.

An accurate detailed description of a low-density gas undergoing shear flow can be obtained whenever the Boltzmann equation is valid. Here we again consider the simpler relaxation-time version of that equation. To start out, we consider a local equilibrium distribution function in which the stream velocity has a gradient,

$$\dot{\epsilon} = du_x/dy :$$

$$f_o(y,v) = (\rho/m)\,(2\pi kT/m)\,e^{-[(mv_x - \dot{\epsilon}y)^2 + mv_y^2]/(2kT)}.$$

If we consider a time-independent field-free system, only the spatial gradient is nonzero and the relaxation-time Boltzmann equation at $y = 0$, where $\langle v_x \rangle$ vanishes, has the form

$$\dot{y}\,\partial f/\partial y = (df/dt)_{collisions} = -f_1 = mv_x v_y \dot{\epsilon} f_o/(kT),$$

where we have approximated $f = f_o + \tau f_1$, and $1/\tau$ is the collision rate. The average pressure-tensor component P_{xy} can then be calculated as an equilibrium average

$$P_{xy} = (\rho/m)\,\langle v_x v_y \rangle_{ne} = -\left[\rho\dot{\epsilon}\tau/(mkT)\right]\,\langle m^2 v_x^2 v_y^2 \rangle_{eq} = -\rho kT\dot{\epsilon}\tau/m = -\eta\dot{\epsilon};$$

$$\eta = \rho kT\tau/m.$$

The viscosity coefficient is just the product of the ideal-gas pressure, $\rho kT/m$, and the collision time τ. This time is about a nanosecond for room-temperature air at atmospheric pressure, and about a thousand times less for a liquid.

The somewhat approximate treatment of the Boltzmann equation can be made exact in the case of two hard spheres, with periodic boundary conditions. In that case, using spherical polar velocity coordinates, the equation of motion reduces to an equation for the motion of a point, representing the hard-sphere velocity, on the surface of a sphere. The surface describes all states of fixed energy for a sphere. This problem has been solved analytically. The solution exhibits shear thinning (decrease of viscosity with increasing strain rate) as well as normal stress effects.

IV.H.4 Numerical Methods.

In 1979 Evans described a straightforward method, with periodic boundary conditions, for solving the many-body equations for an isothermal (isokinetic) molecular dynamics system, using a highly-idealized model of 108 methane molecules undergoing shear. He describes the macroscopic velocity field by the linear function

$$u = u_o + A \cdot r,$$

and determines at each timestep the least-square values of the vector u_o and the tensor A. A new velocity is then chosen for each particle which reproduces the desired flowfield

$$u' = Br; \quad \dot{q} = u + (p/m)$$

and the velocities relative to the local velocity are scaled, in order to reproduce the desired temperature, $T = \langle p^2/(Dmk) \rangle$ in D dimensions.

This accelerating and scaling process can alternatively be carried out in a continuous way by solving the corresponding differential equations of motion. In the simplest case, with $u_x = \dot{\epsilon}y$ the equations of motion are:

$$\dot{x} = (p_x/m) + \dot{\epsilon}y;$$
$$\dot{y} = (p_y/m);$$
$$\dot{p}_x = F_x - \varsigma p_x - \dot{\epsilon}p_y;$$
$$\dot{p}_y = F_y - \varsigma p_y;$$

where the Gaussian isothermal friction coefficient ς has the form

$$\varsigma = \sum \left[F \cdot (p/m) - (\dot{\epsilon}p_x p_y/m) \right]/(2K_o),$$

where K_o is the (constant) value of the kinetic energy.

These shear flow equations can be applied equally well to fluids, to determine the shear viscosity, or to glasses or crystalline solids, to find the yield strength. The extension to polyatomic molecules has also been carried out.

IV.H.5 Fluid Results.

Hundreds of simulations of fluid viscosity with systems ranging from two to a hundred thousand particles have been studied. Many of the data can be correlated with equilibrium thermodynamic properties, as was suggested by Rosenfeld, who plotted a reduced (dimensionless) viscosity as a function of the reduced excess entropy S^e, as shown in **Figure 20**.

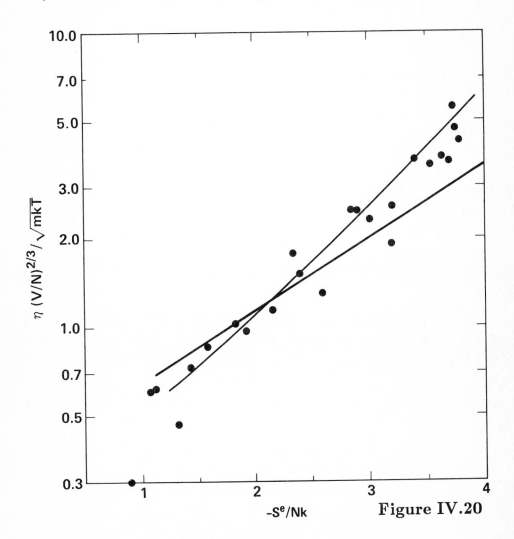

$-S^e/Nk$ **Figure IV.20**

The excess is measured relative to an ideal gas at the same density and temperature. Because Rosenfeld chose *macroscopic* reduction parameters — volume and temperature— for the viscosity rather than *microscopic* potential parameters, his corresponding states relation can be and has been applied directly to real materials. Here the predictions have been compared to a variety of computer simulation results. The straight line best fitting the data predicts fluid viscosities within 30% of the molecular dynamics values for a wide range of force laws.

Why should viscosity be related to entropy? The answer is clear from Andrade's model. Andrade's viscosity varied as the Einstein frequency. The fluid vibrational(!) partition function could be *approximated* by the $3N$ power of a single-particle Einstein-model partition function

$$Z_{vibrational} = \left(kT/h\nu_{Einstein}\right)^{3N} = e^{-3N+(S/k)}.$$

combining this relation, $S/Nk \sim -3\ln\nu$, with the Andrade-Einstein relation, $\ln\eta \sim \ln\nu$, establishes that the Rosenfeld plot *should* be a straight line with a slope of one third. The actual slope shown in **Figure 20** is closer to unity.

Less is known about the nonlinear strain-rate dependence of the viscosity. In two dimensions the viscosity varies as $\ln\dot\epsilon$ for high strain rates and is approximately independent of $\dot\epsilon$ at smaller rates. In three dimensions the data can be described approximately by a linear relation in $\dot\epsilon^{1/2}$. There is also a simple barrier-jumping model due to Eyring which predicts that the shear stress should vary as the inverse hyperbolic sine of the strain rate. No complete analyses of these nonlinear behaviors have appeared yet, but the problem is undergoing intensive investigation.

IV.H.6 Solid Results.

Solids can be deformed too, by applying the same equations of motion Evans used for fluids. In solids the stress rises to a fairly high value, perhaps five percent of the elastic shear modulus. At that point a pair of dislocations is generated somewhere in the crystal, the shear stress drops, and the pair separates to lower the energy. Such a dislocation pair, in a hexagonal cell satisfying periodic boundary conditions, is shown in **Figure 21**. The **Figure** should be viewed obliquely. Compare to the periodic cell containing *three* dislocations shown in Figure 16 of Chapter III.

The motion of a dislocation pair through a crystal in the x direction induces a shear strain of $2b/L_y$. In terms of the plastic strain rate $\dot\epsilon^{plastic}$, the dislocation velocity $v_{dislocation}$, and the dislocation density $2/L_xL_y = 2/V$, Burgers' relation results

$$\dot\epsilon^{plastic} = (Nv)_{dislocation}\, b/V.$$

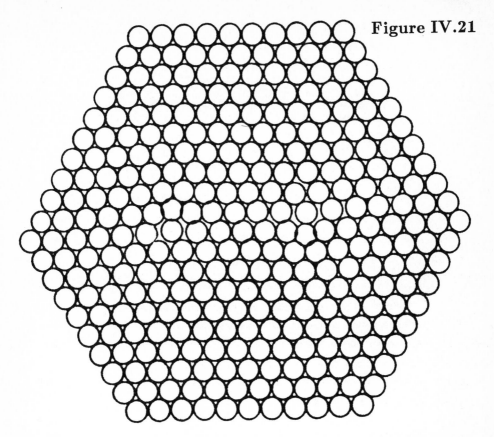

Figure IV.21

The dislocations can be clearly seen in movies made with molecular dynamics simulations, but their behavior is a little complicated. The usual "theory" sets the rate of dislocation production (proportional to the jump rate over an energy barrier) equal to the rate of annihilation, which varies as the square of the dislocation density. Molecular dynamics simulations show instead that the production rate varies as the *cube* of the dislocation density, leading to the straight lines shown in the double logarithmic plot of a dimensionless shear stress, σ/G as a function of the dimensionless strain rate $(\dot{\epsilon}d/c)$ shown in **Figure 22**. G, d, and c are respectively the shear modulus, interparticle spacing, and transverse sound velocity. By using reduced stress and reduced strainrate, results from computer calculations in both two and three dimensions can be compared with data, either directly measured or inferred, for real materials. The data in **Figure 22** were obtained by analyzing the shapes of plastic deformation waves in several metals. The molecular dynamics results shown in the **Figure** include both two- and three-dimensional systems. These two kinds of molecular dynamics results agree within the width of the line.

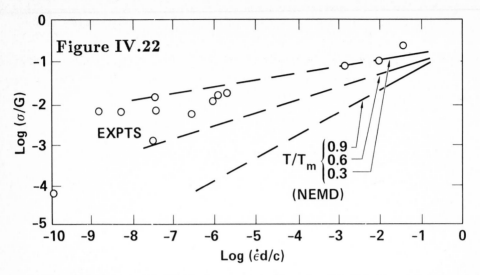

Figure IV.22

The relatively untouched field of *mesoscopic* dynamics, in which the particles are inter-mediate in scale between atoms and macroscopic flow features, appears relatively promising for solid plasticity. In such simulations the individual particles are dislocations in two dimensions, or elements of dislocation lines in three dimensions, which interact with a tensor force and which can be created and destroyed with rate laws depending upon the local stress and temperature. The few simulations that have been carried out in this way are nicely consistent with the much more expensive atomistic simulations.

IV.I Simulation of Heat Flows

IV.I.1 Fourier's Law

Heat obeys the diffusion equation too, with an effective diffusion coefficient called the "thermal diffusivity". Fourier's linear phenomenological law,

$$Q = -\kappa \nabla T,$$

relates the heat flux Q to the thermal conductivity κ and the temperature gradient ∇T. Like Newtonian viscosity, the basic equation defining conductivity is irreversible. Q is odd in velocity and ∇T even. Fourier's linear relation leads to the diffusion equation if it is assumed that the conductivity κ is a constant and if the temperature increase in the material is proportional to the rate at which it gains heat, $-\nabla \cdot Q$. Then the thermal diffusivity D_T is the ratio of κ to the constant-pressure heat capacity, per unit mass, C_P, times the mass density:

$$(\partial T/\partial t) = D_T \nabla^2 T; \quad D_T = \kappa/(\rho C_P).$$

It is often convenient to solve heat-flow or diffusion problems using a Green's function solution (a Gaussian which spreads as the square root of the time) or a Fourier series. The Gaussian solution of the diffusion equation was displayed in Section G.2 of this Chapter. If we follow the Fourier series approach we express the temperature at any time t in terms of the Fourier amplitudes $A_k(t)$, so that

$$T(r,t) = \sum A_k(t)\, e^{ik \cdot r}.$$

Then the diffusion equation gives a simple exponential decay for each amplitude.

$$A_k(t) = e^{-k^2 D_T t} A_k (t = 0).$$

This relation gives useful order-of-magnitude estimates for the time required for thermal equilibration. Just as in the case of ordinary diffusive and viscous flows, the phenomenological coefficient—here D_T—is of order [centimeters^2/seconds] for gases, and some *five* orders of magnitude smaller for insulating solids and liquids.

IV.I.2 Irreversible Thermodynamics

The mixing caused by diffusion and the viscous dissipation caused by stirring are easy to visualize. Mechanical work is converted into heat. The dissipation caused by heat flow is not so closely linked to physical experience, but, from thermodynamics, the work which can be obtained from a given quantity of heat depends upon the temperature, so that as heat is transferred from a hot to a cold body the capacity to do work is diminished. The dissipation or "lost work" is expressed in terms of the rate of change of the thermodynamic entropy S where $\dot{S}$ is $(dQ/dt)/T$ for a reversible heat-transfer process. Bear with the circumstance that Q, in thermodynamics, is an *amount* of heat, as opposed to the *flux* of heat.

Consider the case of a two-dimensional bar mentioned in Section D of this Chapter, shown in **Figure 23**, connected to two ideal-gas heat reservoirs at its ends, at temperatures T_H and T_C.

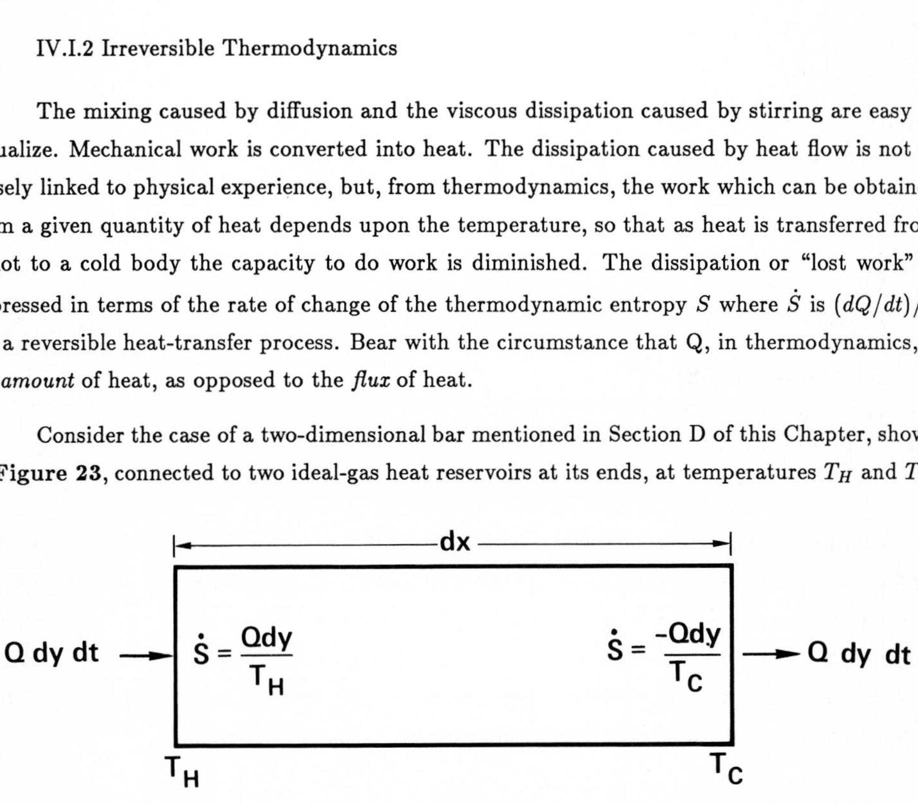

Figure IV.23

As the hot reservoir transfers heat to the bar, at a rate $Q\,dy$, the reservoir loses and the bar gains entropy at a rate $Q\,dy/T_H$. A cooler reservoir at the other end gains entropy $Q\,dy/T_C$. The sum of the two rates for the bar, one at the hot end and one at the cold end, is *negative*! But in the steady state it is clear that the bar's entropy must instead be constant. This can only be the case if the entropy gain and loss, at the ends of the bar,

$$\dot{S}_{ends} = Q\,dy\left[(1/T_H) - (1/T_C)\right],$$

is supplemented by an irreversible entropy production,

$$\dot{S}_{internal} = (Q\,dy/T)\left[(T/T_C) - (T/T_H)\right] = -(Q\,dy/T)(1/T)\langle\,dT/dx\,\rangle\,dx.$$

so that the total steady rate of entropy change is zero:

$$\dot{S}_{bar} = \dot{S}_{ends} + \dot{S}_{internal} = 0.$$

In the "linear" regime where Fourier's law holds, we can expand the high and low temperatures about the average value and replace the temperature gradient with $-Q/\kappa$ with the result

$$T\dot{S}_{internal} = Q_x^2 V/(\kappa T).$$

This result resembles those for diffusion and viscosity. In all three cases the rate at which (free) energy is converted into (lost) heat varies as the square of the corresponding flux.

IV.I.3 Einstein Conductivity Model of Horrocks and McLaughlin

Horrocks and McLaughlin suggested that the heat conductivity could be estimated from an Einstein model resembling Andrade's model for viscosity just described in Section H. The idea is the same. A particle oscillating at the Einstein frequency $\nu_{Einstein}$ will transport *heat* from its hotter to its cooler neighbors, through a microscopic area of order d^2. If we use the classical duLong-Petit heat capacity, $3k$ per atom, then the transfer of energy per unit time and area is approximately $-3kd\,(dT/dx)\,\nu/d^2$. For this to reproduce Fourier's linear relation, $Q = -\kappa\,\nabla T$, the conductivity κ must be given by the relation

$$\kappa = 3k\,\nu_{Einstein}/d.$$

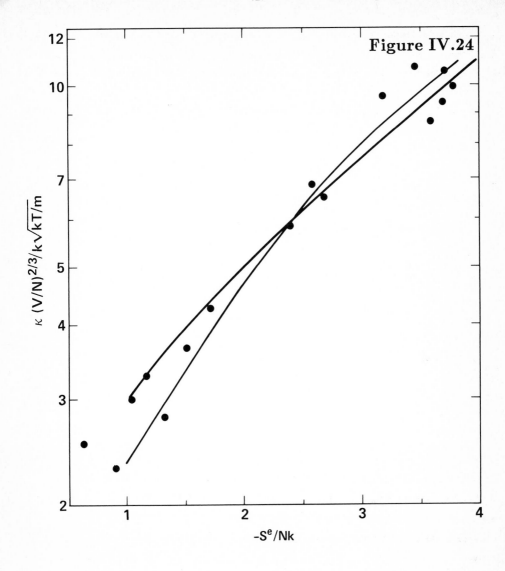

Figure IV.24

This crude model is considerably better than might be expected. In **Figure 24** simulation data are displayed as points and curves, again using macroscopic parameters, volume and temperature, to define a reduced transport coefficient. Again S^e is the excess entropy. The slope of the plot is somewhat greater than the prediction of 1/3 which follows from Horrocks and McLaughlin's model. The more sophisticated phonon scattering theory predicts that the conductivity varies as $1/T$. For dense fluids the conductivity is considerably closer to Horrocks and McLaughlin's temperature-independent relation.

IV.I.4 Boltzmann Equation

An approximate Boltzmann equation can be written for the heat-flow problem. To do so, consider an unperturbed distribution function that has density and temperature gradients satisfying the constant-pressure relation

$$(d\ln\rho/dx) + (d\ln T/dx) = 0.$$

Unless the gradients balance in this way the resulting pressure gradient would generate sound or shock waves to smooth out the force imbalance.

In the low-density Boltzmann-Equation regime, the unperturbed distribution has the form

$$f_o = \left[\rho(x)/m\right] e^{-mv^2/[2kT(x)]} \left[2\pi kT(x)/m\right]^{-D/2}.$$

The spatial gradient terms in the Boltzmann equation operate on both the density and the temperature. But these variables are constrained to obey the constant-pressure ideal gas law, with the product ρT constant. The familiar procedure of expanding the solution of the relaxation-time Boltzmann equation about f_o leads to a perturbation expansion:

$$f = f_o + f_1\tau + \ldots \; ;$$

with

$$f_1 = -f_o\dot{x}\left(d\ln T/dx\right)\left[(mv^2/2kT) - (D/2) - 1)\right]$$

from which the conductivity can be estimated. In Boltzmann's rigorous kinetic theory, in which the dependence of collision time on velocity is correctly taken into account, the heat conductivity is given, within about 1%, by the same cross section integral as is the shear viscosity. On the other hand, the heat flux seems to be intrinsically more complicated than viscous momentum flux, involving a higher moment of the velocity distribution and having a vanishing value at equilibrium. Another example of this relative complexity is that there is no heat flux in a two-body system with fixed center of mass. Symmetry requires that the two particles move oppositely so that the net heat flow must vanish. Three particles are enough to generate a heat current, as we will emphasize in the next section. But three particles lie outside the scope of the two-particle Boltzmann-Equation description, which is concerned with the effect of uncorrelated binary collisions. Thus the close connection found at low density between viscous flow and heat flow seems to have no parallel for liquids. Viscous flows are most simply generated by moving boundaries which incorporate the strain rate. Heat flows use instead an external force, first derived by Evans and Gillan, to drive the heat current Q.

IV.I.5 Evans and Gillan's Driving Force for Heat Flow

Evans and Gillan independently showed how to relate the dissipation from heat flow to an external force in such a way as to reproduce the Green-Kubo heat conductivity,

$$\kappa = (V/kT^2) \int_o^\infty \langle Q_x(0)Q_x(t) \rangle \, dt.$$

Their *styles* are very different. Gillan uses approximately 100 equations in his derivation; Evans uses 3. But the results are essentially the same: to generate a heat current flowing in the x direction, the additional force

$$\lambda(\delta E + V\delta P_{xx}^\phi, \, V\delta P_{xy}^\phi, \, V\delta P_{xz}^\phi)$$

is added to the equations of motion. This force reproduces the Green-Kubo heat-flux autocorrelation formula for κ and the thermodynamic dissipation $T\dot{S} = Q_x^2 V/(\kappa T)$.

This nonequilibrium force is the basis for an efficient simulation method generating conductivities using periodic boundary conditions. A variety of solid and fluid systems, ranging from several hundred Lennard-Jones particles down to three hard disks, have been studied.

The heat flow problem is unique in that it provides the only known mechanical problem in which Gauss' Principle of Least Constraint, described in Section D of Chapter I, predicts *incorrect* results. If that principle is used to constrain the heat flux to a constant value, equations of motion slightly different from those of Evans and Gillan result. In the Gauss' Principle version it is the complete pressure P, not just the potential part P^ϕ, which couples to the external field λ. It is possible to show that this produces erroneous results, in conflict with the Green- Kubo expression for heat conductivity. The numerical consequences of this error have been shown to be small, of order 10% or less. This case is unique. The other nonequilibrium methods we have outlined, for diffusion and viscosity as well as heat flow, all can be shown to reproduce the exact Green-Kubo linear-response results.

IV.I.6 Results

The conductivities found for fluids are displayed in **Figure 24** in the simple way suggested by the Horrocks and McLaughlin model. In that model the conductivity is proportional to the Einstein frequency. The Einstein model can be used to estimate thermodynamic properties too, from the approximate partition function

$$e^{-A_{Einstein}/(NkT)} = \left[kT/(h\nu_{Einstein})\right]^3 = e^{-3+(S/Nk)}.$$

This partition function suggests that the entropy per particle varies as $-3k\ln\nu_{Einstein}$, so that a semilogarithmic plot of logarithm (conductivity) versus entropy should be a straight line, with slope one-third. Although the best slope for such a line is actually about 0.4 rather than 0.33, the basic idea is an extremely useful one. There is an excellent correlation between the computer simulation results for a wide range of force laws linking the conductivity to the entropy.

Heat flow requires three particles and, for a finite conductivity, a mechanism for dissipation. This dissipation is the scattering of energy travelling in the x direction into the y direction. The simplest system meeting both requirements is three hard disks. There is no difficulty in solving the isothermal or isoenergetic equations of motion for three disks. The main surprise is that the pressure and heat flux in such a system differ for the two kinds of mechanics. The fluxes are somewhat greater in the isokinetic case. This difference, between the isokinetic and isoenergetic fluxes, emphasizes the point that hard disks and spheres only exist as idealizations.

The resulting heat conductivities for the three particles lie well below those expected for a large system at the same density and temperature. But the conductivities vary nearly linearly with the logarithm of the driving force λ over the 32−fold range of values studied. See **Figure 25**. This logarithmic behavior is predicted by the mode-coupling hydrodynamic theory. It is surprising to find the *same* dependence in a very small system with only a three-dimensional velocity space, six velocity components less two for the center of mass and one for the energy conditions. This result, and similar results for viscous flows and for three-dimensional flows, suggest that the nonlinear behavior of systems far from equilibrium can probably be understood qualitatively on the basis of very small system results.

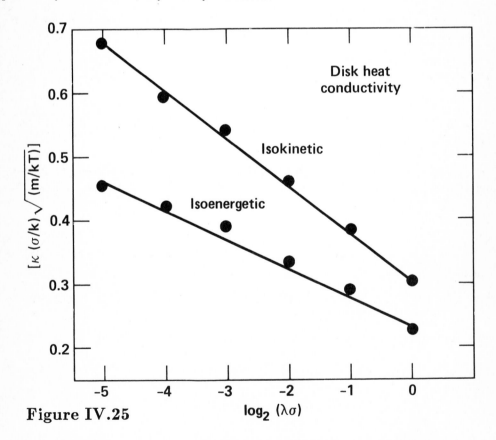

Figure IV.25

Bibliography for Chapter IV.

Nonequilibrium methods are reviewed by D. J. Evans and W. G. Hoover, "Flows Far From Equilibrium *via* Molecular Dynamics", Annual Reviews of Fluid Mechanics **18**, 243 (1986). This work contains a comparison of the various thermostats suggested by control theory.

The Information-Theory approach of Jaynes and Zubarev is discussed in W. G. Hoover, "Nonlinear Conductivity and Entropy in the Two-Body Boltzmann Gas", Journal of Statistical Physics **42**, 587 (1986).

B. B. Mandelbrot, *The Fractal Geometry of Nature* (W. H. Freeman, San Francisco, 1982) is a readable introduction to a fascinating subject. The several slightly different definitions of fractal dimensionality are carefully contrasted and applied by J. D. Farmer, E. Ott, and J. A. Yorke, in "The Dimension of Chaotic Attractors", Physica **7D**, 153 (1983).

Gauss' Principle is applied to many-body diffusion in D. J. Evans, W. G. Hoover, B. H. Failor, B. Moran, and A. J. C. Ladd, "Nonequilibrium Molecular Dynamics *via* Gauss' Principle of Least Constraint", Physical Review **28A**, 1016 (1983).

The simulation of dense-fluid bulk viscosity through the hysteresis associated with cyclic compression and dilation is described in W. G. Hoover, A. J. C. Ladd, R. B. Hickman, and B. L. Holian, "Bulk Viscosity *via* Nonequilibrium and Equilibrium Molecular Dynamics", Physical Review **21A**, 1756 (1980).

The algorithm for shear viscosity is described in D. J. Evans, "Nonequilibrium Molecular Dynamics Study of the Rheological Properties of Diatomic Liquids", Molecular Physics **42**, 1355 (1981).

Nonequilibrium corresponding states relationships are described by Y. Rosenfeld, "Relation Between the Transport Coefficients and the Internal Entropy of Simple Systems", Physical Review **15A**, 2545 (1977).

The mesoscopic flows of dislocations are described in A. J. C. Ladd, "Molecular Dynamics Studies of Plastic Flow at High Strain Rates", Topical Conference on Shockwaves, Spokane, Washington (July, 1985).

The Evans-Gillan method for generating heat flows without temperature gradients is described in D. J. Evans, "Homogeneous Nonequilibrium Molecular Dynamics Algorithm for Thermal Conductivity–Application of Noncanonical Linear Response Theory", Physics Letters **91A**, 457 (1982). A series of simulations based on the method, as described in this Chapter, appear in W. G. Hoover, B. Moran, and J. M. Haile, "Homogeneous Periodic Heat Flow *via* Nonequilibrium Molecular Dynamics", Journal of Statistical Physics **37**, 109 (1984).

V. Future Work

A decade ago Mansoori and Canfield suggested perturbation methods which led to quantitative solutions of equilibrium problems for simple dense fluids. Now, methods are being developed to simplify the treatment of nonequilibrium problems. There already exist efficient trustworthy molecular dynamics algorithms for simulating diffusive, viscous, and conducting flows as well as flows coupling pairs of these phenomena. *All* of these schemes are consistent with the Green-Kubo linear response theory, and *all* of these schemes produce reasonable nonequilibrium flows, although *none* of them has been proven valid in the nonlinear regime. Much work remains to be done in articulating the "well-posed problems" needed to justify the nonequilibrium methods.

From the standpoint of applications to materials science and engineering the slow progress in solving the quantum many-body problem guarantees that inverse power potentials, as well as combinations such as the Lennard-Jones 6-12, and exponential-six potentials, and the more intricate semi-theoretical models will dominate molecular dynamics simulations, for the forseeable future.

In problems involving homogeneous flows, without boundaries, the nonequilibrium algorithms have reduced number-dependence in the results. But there are many important problems involving physical gradients, such as shock and detonation waves in which *all* the field variables cover a range of values. For such inhomogeneous systems larger and faster simulations are necessary. Because individual processors are reaching limiting speeds, improvements are being mainly obtained by combining many processors in parallel. Simulations with millions of particles will soon be a reality. Likewise, the possibility of storing and processing *billions* of numbers makes it possible to characterize distribution functions in quantum problems and nonequilibrium problems, areas now made difficult simply by the requirements on storage capacity.

The flexibility in simulations will make it possible to follow the flow of relatively complicated molecules in channels, to solve problems involving friction and wear, potential energy surfaces with chemical reactions, and quantum mechanics. In the quantum case it is not yet clear how to proceed in problems for which no Hamiltonian is available. It is for this reason that the Nosé Hamiltonian thermostats will prove particularly useful. When these thermostats are applied to quantum systems even present computer capacity is sufficient to treat two- or three-body problems, such as those discussed in Sections D and F of Chapter IV.

Recent developments in the theory of dynamical systems have revealed shortcomings in the traditional theoretical approach to statistical mechanics, which ignores the mixing effect of periodic boundaries as well as the Kolmogorov entropy associated with Lyapunov instability. The language and computational techniques borrowed from nonlinear dynamics will aid the development of the nonequilibrium theory. An old approach which also promises to gain from the new techniques is the dynamical response method developed by Gianni Iacucci, Giovanni Ciccotti, and Ian McDonald.

Thus we anticipate that the mysteries of entropy far from equilibrium, the problem which motivated Boltzmann, will soon be effectively laid to rest by quantitative simulations. The work that Boltzmann began is accelerating now and shows no sign of reaching limits on its ability to stimulate the development of the sound physical theory necessary to the analysis and control of physical processes far from equilibrium.

Bibliography for Chapter V.

For a look ahead there is no better source than the Proceedings of recent conferences. See "Nonlinear Fluid Behavior", the Proceedings of the Boulder, Colorado 1982 Conference, published in Physica **118A**, 1 (1983) and the Proceedings of the Lago di Como 1985 Enrico Fermi International Summer School, "Molecular Dynamics Simulation of Statistical-Mechanical Systems", soon to be published by North-Holland Publishing Company and the Italian Physical Society.

INDEX

Adiabatic Sound Waves 60

Andrade's Viscosity Model 118,122

Appell's Cart 23

Barrier Jumping *via* Lagrange's Mechanics 13

Bill Ashurst 92

Bird's Approximation 75

Boltzmann Equation 72

Boltzmann Equation for Diffusion 103,104,111

Boltzmann Equation for Heat Flow 128

Boltzmann Equation for Viscosity 119

Bouncing Ball and Lyapunov Instability 8,9

Canonical Probability Density 17,29

Cantor Set 107

Coexisting Phases 69

Comoving Frame 47

Compressible Flow of Phase-Space Probability 29

Conductivity for Fluids 129

Conductivity for Three Hard Disks 100,130

Constraint Forces 92

Continuity Equation of Continuum Mechanics 48

Continuum Mechanics for Shockwaves 82

Control Theory 94

Constant-Temperature (Gaussian) Mechanics 26

Constant-Temperature (Nosé) Mechanics 19

Corresponding States for Heat Conductivity 127

Corresponding States for Viscosity 121

Crack Healing 88

Crack Inertia 88

Damped Oscillators 28

Diatomic Constraint Force 24

Differential Control 94,99

Diffusion Coefficient for Two Hard Disks 104

Diffusion Equation Solution from Fourier Series 125

Dislocation Energy 89

Dislocations in Solid Plasticity 122

Displacement Coordinates for Elasticity Calculations 57

Driving Forces 92

Einstein Conductivity Model of Horrocks and McLaughlin 126

Einstein Viscosity Model of Andrade 118

Elastic Constants 59

Elasticity 114

Elastic Strains 57

Energy Equation of Continuum Mechanics 52

Enskog's Viscosity Model 118

Entropic Variables 43

Equation of Motion of Continuum Mechanics 50

Ergodicity 4,45

Eulerian Coordinates 47

Evans-Gillan Heat-Current Driving Force 98,129

Eyring's Viscosity Model 122

Feedback 94

Fermi-Pasta-Ulam One-Dimensional Chain 35

Fick's First and Second Laws of Diffusion 108

First Law of Thermodynamics 52

Force Law for Rare Gases 72

Fourier Heat Conduction and Thermal Diffusivity 124

Fractals 106

Fragmentation Simulation 84

Free Volume and Chemical Potential 45

Friction Coefficient ς 31,45

Future Developments 132

Galton Board 102

Gauss' Mechanics 22

Gauss' Principle 22,24

Gauss' Principle Failure for Heat Flow 129

Generalized (Phase-Space) Continuity Equation 27

Gibbs' Variational Principle 69

Grady-Glenn Model for Fragmentation 83

Gravitational Forces and the Thermodynamic Limit 11

Gravity 1

Green-Kubo Linear-Response Theory 98,101

Hamilton's Mechanics 16

Harmonic Oscillator 19

Heat Conducting Chain 97

Heat Conductivity for Three Hard Disks 100,130

Heat-Current Driving Force 99

Heat Flux Vector 52

Heat Theorem 54,59

Heisenberg Picture 47

Holonomic Constraints 13

Homogeneous Deformation 57

Hugoniot 80

Information Theory 44

Information Theory Pitfall 105

Integral Control 96,99

Interface Between Fluid and Solid 92

Irreversibility 73,110

Irreversible Thermodynamics of Heat Flow 125

Isokinetic Canonical Distribution 29

Isothermal Constraint Forces 94

Isothermal Molecular Dynamics 94

Jaynes' Information Theory 44,105

Jaynes-Zubarev Distribution Functions 101

Kepler's Laws 1

Kolmogorov-Arnold-Moser Tori 35

Lagrange Multiplier 25

Lagrange's Mechanics 13

Lagrangian (Comoving) Coordinates 47

Lamé Elastic Constants 115

Langevin Equation 95

Lattice-Coordinates Virial Theorem 56

Least Action Integral 14

Limit Cycles (Rayleigh and van der Pol Equations) 96

Liouville Equation 31

Loschmidt Objection 4

Lyapunov Instability 6

Mayers' Expansion of Dense Gas Partition Function 62

Mean-Squared-Displacement Divergence in Two Dimensions 63

Mechanical Variables 43

Melting of Hard Hyperspheres 65

Melting of Two Hard Disks 65,66

Mesoscopic Dynamics of Dislocation Motion 124

Mixing in Phase Space from Periodic Boundaries 34

Newtonian Molecular Dynamics 72

Newton's Mechanics 1

Nonholonomic Constraints 22

Nonlinear Diffusion Coefficient from Molecular Dynamics 113
Nonlinear Nonholonomic Constraints 23
Nonlinear Response Theory 101
Nosé Hamiltonian 20
Nosé Ideal Gas 32
Nosé Mechanics Derivations 31
Nosé Oscillator 20
Nosé's Mechanics 16,27
Number-Dependence 62
One-Dimensional Chain 66
Periodic Boundary Conditions 10,63
Phase Diagram for Hard Spheres 68
Phase-Space Mixing 34
Phonon Scattering Theory of Heat Conductivity 127
Phonons in a Three-Particle Chain 98
Plane Couette Flow 115
Plasticity 114
Poincaré Section for Nosé Oscillator 35
Power Loss 99
Pressure Tensor 50,58
Pressure Tensor Symmetry 51
Principle of Least Constraint 22
Probability Density in N-Body System 27
Proportional Control 95,99
Puncture Plot 35
Radiation Damage Simulations 39
Rankine-Hugoniot Equation 80
Rayleigh Line for Shockwave States 80
Rayleigh's Equation for Nonlinear Oscillations 95
Relaxation-Time Boltzmann Equation 74,111,119,128
Reversibility 4
Reversibility Paradox 35,73
Reynolds Number 117
Rigid Diatomic Molecule using Gauss' Principle 24
Rigor Mortis School of Continuum Mechanics 115
Rigor Mortis School of Statistical Mechanics 6
Rigorous Kinetic Theory 128
Rosenfeld Relation for Viscosity 121
Schrödinger Picture 47

Second Law of Thermodynamics 4,37
Shear Deformation 85
Shear Thinning for Two Hard Disks 106
Shockwave Kurtosis 82
Shockwave Profile 78
Shockwave Simulation 77
Shockwave Width 81
Sierpinski Sponge 107
Stable Motion 7
State Variables 42
Steady-State Atmosphere from Boltzmann's Equation 111
Strain-Rate Tensor 114,115
Stress Concentration 87
Symmetry of Pressure Tensor 51
Temperature 27,43
Temperature-Dependent Hamiltonian of Nosé 19
Temperature of a Many-Body System 26
Thermodynamic Limit 12,117
Three-Particle Harmonic Chain 67,97
Time-Dependent Perturbation Theory 60
Time Reversal 5,19
Time Scaling of Nosé 21
Toda Potential for Ideal Gas Energy 32
Triangular-Lattice Frequency Distribution 86
Triatomic Molecule with Two Holonomic Constraints 14
Triple Point Simulation 69
Two-Particle Relaxation-Time Boltzmann Equation 74
Uncertainty Principle 9
Unstable Motion 8,9
van der Pol's Equation from Rayleigh's Equation 95
Verlet (Störmer) Algorithm 3
Virial Theorem 2,54
Viscoelasticity 61
Viscosity 114,121,122
Viscosity for Two Disks 65
Viscous Hysteresis 116
Vortex Shedding 75
Vorticity Tensor 115
Zermelo-Poincaré Objection 4